Flame-Retardant Polymeric Materials

Volume 3

CONTRIBUTORS

Joseph Green
Chemical Research and Development Center
FMC Corporation
Princeton, New Jersey

Stanely C. Israel
Polymer Science Program
Department of Chemistry
University of Lowell
Lowell, Massachusetts

John F. Krasny
Center for Fire Research
National Bureau of Standards
Washington, D.C.

David F. Lawson
Central Research Laboratories
Firestone Tire and Rubber Company
Akron, Ohio

Archibald Tewarson
Factory Mutual Research Corporation
Factory Mutual System
Norwood, Massachusetts

Flame-Retardant Polymeric Materials

Volume 3

Edited by

Menachem Lewin
Israel Fiber Institute
Jerusalem, Israel

S. M. Atlas
Bronx Community College of CUNY
Bronx, New York

and

Eli M. Pearce
Polytechnic Institute of New York
Brooklyn, New York

PLENUM PRESS · NEW YORK AND LONDON

Library of Congress Cataloging in Publication Data

Main entry under title:

Flame-retardant polymeric materials.

Includes bibliographical references and indexes.

1. Fire resistant polymers. I. Lewin, Menachem. II. Atlas, Sheldon M. III. Pearce, Eli M.

TH1074.F58 668.9 75-26781
AACR1

ISBN 978-1-4757-0114-2 ISBN 978-1-4757-0112-8 (eBook)
DOI 10.1007/978-1-4757-0112-8

A Division of Plenum Publishing Corporation
233 Spring Street, New York, N.Y. 10013
Softcover reprint of the hardcover 1st edition 1982

Preface

Flammability has been recognized as an increasingly important social and scientific problem. Fire statistics in the United States (Report on the National Commission on Fire Prevention and Control, "America Burning," 1973) emphasized the vast devastation to life and property—12,000 lives lost annually due to fire and these deaths are usually caused by inhaling smoke or toxic gases; 300,000 fire injuries; 11.4 billion dollars in fire cost of which 2.7 billion dollars is related to property loss; a billion dollars to burn injury treatment; and 3.3 billion dollars in productivity loss. It is obvious that much human and economic misery can be attributed to fire situations. In relation to this, polymer flammability has been recognized as an increasingly important social and scientific problem. The development of flame-retardant polymeric materials is a current example where the initiative for major scientific and technological developments is motivated by sociological pressure and legislation. This is part of the important trend toward a safer environment and sets a pattern for future example. Flame retardancy deals with our basic everyday life situations—housing, work areas, transportation, clothing and so forth—the "macroenvironment" capsule within which "homosapiens" live. As a result, flame-retardant polymers are now emerging as a specific class of materials leading to new and diversified scientific and technological ventures.

From the humble beginnings of flame-retardance treatments of existing polymers, the field is now developing into the design and engineering of new heat-resistant molecules, polymers, and commodities which are inherently flame-retardant. It is an interdisciplinary development and involves several scientific, engineering, legal, medical, and sociological consequences.

The objective of this series is varied. In many cases it is to give an up-to-date summary of the state of the art in flame-retarding polymeric materials so as to be an aid to those involved in solving these problems. Interpretation as to mechanism and conjecture about future approaches has been encouraged on the part of the authors. Since polymer degradation is the precursor to flammability, suitable importance has also been placed on this area. There will be occasional chapters which also deal with a specific test method, some of which may have historical importance.

We are hopeful that through these means, meaningful solutions to a number of the flammability problems and their subsequent positive social ramifications will be accomplished.

M. Lewin
S. M. Atlas
E. M. Pearce

Contents

CHAPTER 3

Experimental Evaluation of Flammability Parameters of Polymeric Materials

ARCHIBALD TEWARSON

CHAPTER 4

Flammability Evaluation Methods for Textiles

JOHN F. KRANSY

1

The Flame Retardation of Polyolefins

Joseph Green

1. Introduction

Polyolefins are flammable and will burn in air with a very hot and clean flame accompanied by melting and dripping like a candle. Essentially no soot is developed in the flame, as normally obtained during the burning of aromatic polymers, and little to no residual char is formed. The present state-of-the-art in polyolefin flame retardation dictates the use of halogen-containing compounds whose effectiveness is enhanced by the use of antimony oxide as a synergist.

It is generally believed that the effectiveness of halogen as a flame retardant for polymers is

$$I > Br > Cl > Fl$$

Fluorocarbon compounds are very poor flame-retardant additives. Iodine compounds are generally believed to be the most effective; however, their thermal stability is too low for practical use in plastics. Both bromine- and chlorine-containing compounds are used commercially as flame-retardant additives. It is generally believed that aliphatic halogen is more effective than

Joseph Green • Chemical Research and Development Center, FMC Corporation, Princeton, New Jersey.

aromatic halogen and the effectiveness of alicyclic halogen falls somewhere in between.

The mechanism of flame retardation of polyolefins had been reviewed previously.[1] The objective of this treatise is to present application data for the use of compounders and end users. The effect of various types of halogen in each polyolefin is compared so that valid generalizations can be made. The polyolefins discussed are polypropylene homopolymer, polypropylene copolymer, low-density polyethylene, high-density polyethylene, cross-linked polyethylene, ethylene–propylene–dimer rubber, and some miscellaneous hydrocarbon polymers.

The chlorine- and bromine-containing compounds of various structures discussed in this paper are shown in Table 1, along with some key properties for comparative purposes. The melting point and thermal stability of these products are the key to their application and will be discussed for each polymer system. The processing temperatures for polyolefins are high (200–300°C) and flame-retardant compounds must be sufficiently stable thermally under mechanical work for practical application.

Only two chlorinated products are used commercially to flame retard polyolefins, the chlorinated paraffins and Dechlorane Plus. As can be seen in Table 1, the chlorinated paraffins have a very low thermal stability and their use is generally limited to flame-retarding blown polyethylene film. Here the processing temperatures reach only about 190°C. Dechlorane Plus has much

TABLE 1
Properties of Typical Flame-Retardant Additives Used in Polyolefin Resins

	% Cl	% Br	Halogen type	Melting point (°C)	Thermal stability (°C)
Chlorinated paraffin	70		Aliphatic	102	~190
Dechlorane Plus	65		Alicyclic	>350	>350
Hexabromocyclododecane		74	Alicyclic	174–186	~175
Tetrabromocyclooctane		74	Alicyclic	105–112	~140
Dibromoethyldibromo-cyclohexane		74	Alicyclic	65–80	~225
Decabromodiphenyl oxide		83	Aromatic	304	330
Saytex BC-26	29	40	Alicyclic	200	260
Saytex BN-21		40	Alicyclic	~300	~300
Saytex BN-451		46	Alicyclic	~300	~300
Saytex BT-93		66	Aromatic	450	450
Bromkal 66-8[a]		65–67	Aliphatic-aromatic	85–100	–
Bromkal 75-5PE[b]		73–75	Aliphatic-aromatic	~40	–

[a]2,2-Bis[4-(2,3-dibromopropoxy)-3,5-dibromophenyl]propane.
[b]2,4,6-Tribromophenoxy-2,3-dibromopropane.

greater thermal stability and is used both in polypropylene and polyethylene. Its effectiveness as a flame retardant, however, is quite poor as compared to brominated compounds. Of the three brominated alicyclic products shown, only dibromoethyldibromocyclohexane possesses adequate thermal stability for application in some polyolefin processing; injection–molded compositions are possible with this compound. Hexabromocyclododecane and tetrabromocyclooctane have low thermal stability and are used only in expandable polystyrene bead board. The two aromatic bromine-containing compounds that are used in polyolefins and discussed in this work are decabromodiphenyl oxide and SAYTEX BT-93; both have very high thermal stability. The two Bromkal derivatives are products made by Kalk/Europe and are used to flame-retard polypropylene pipe. Bromkal 73-5PE has a very high vapor pressure and a very bad odor during processing. The next three products, SAYTEX BC-26, BN-21, and BN-451, are highly effective alicyclic bromine-containing compounds useful for a variety of applications in polypropylene homopolymer and copolymer. Their high thermal stability allows them to be used in high-temperature processing applications and their major application is for UL-94 V-2 compositions. The commercial uses of these compounds will be discussed in detail in subsequent sections. The thermal characterization of the flame retardants listed in Table 1 by thermogravimetric analysis (TGA), differential thermal analysis (DTA), and differential scanning calorimetry (DSC) is shown in Fig. 1a-v.

Oxygen index data are used as a measure of flame retardancy because of the highly reproducible nature of this test. The flame retardancy is also checked by the UL-94 vertical burn test. Both of these tests are useful in comparing the flame-retardant effectiveness of different types of compounds in the laboratory. These data, however, cannot be used to predict performance or assess hazard under actual fire conditions. Terms such as self-extinguishing

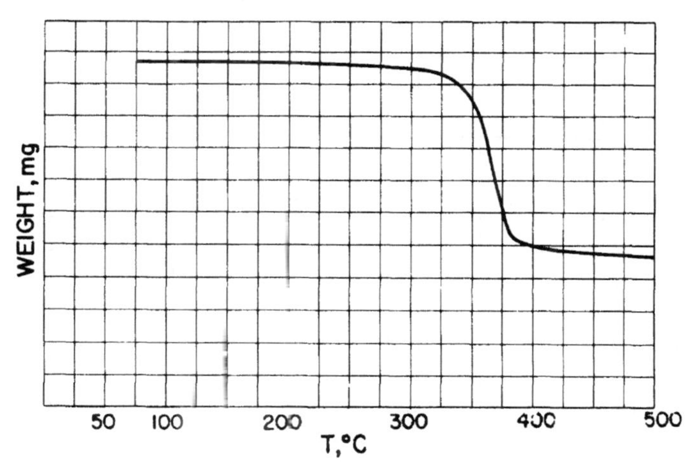

FIGURE 1a. Thermogravimetric analysis (TGA) Chlorowax 70.

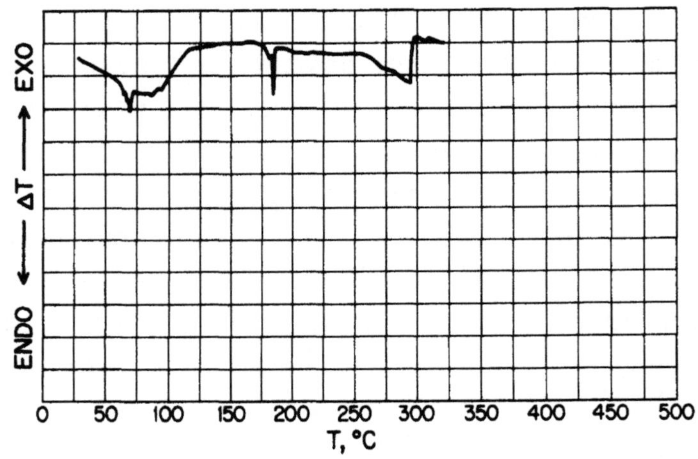

FIGURE 1b. Differential thermal analysis (DTA) Chlorowax 70S.

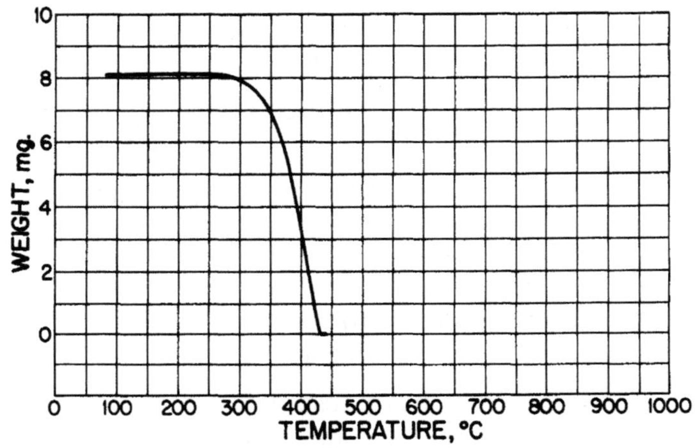

FIGURE 1c. Thermogravimetric analysis (TGA) Dechlorane Plus 515.

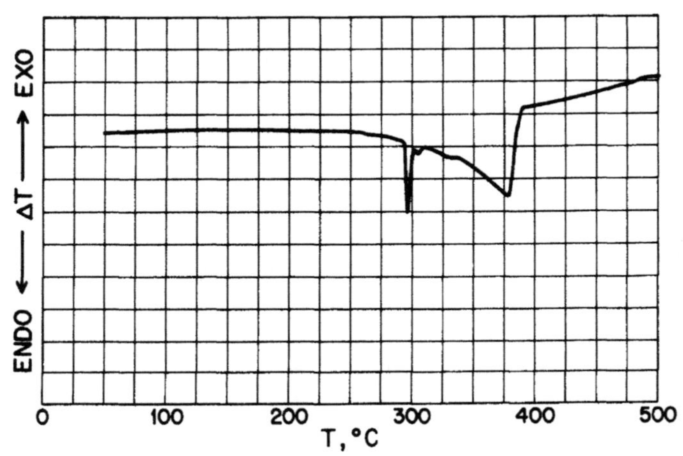

FIGURE 1d. Differential scanning calorimetry (DSC) Dechlorane Plus 515.

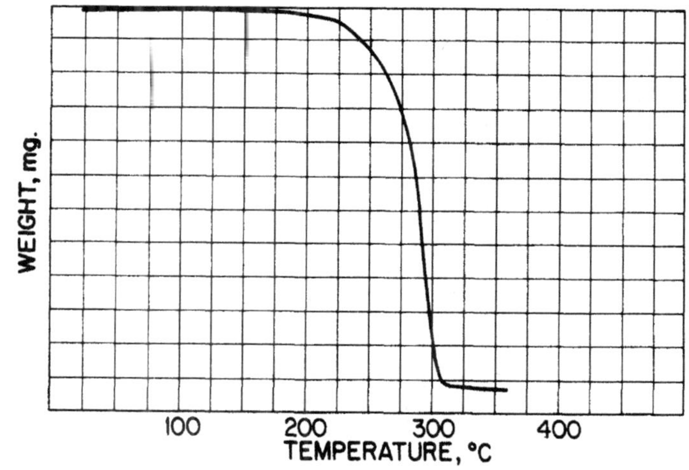

FIGURE 1e. Thermogravimetric analysis (TGA) hexabromocyclododecane.

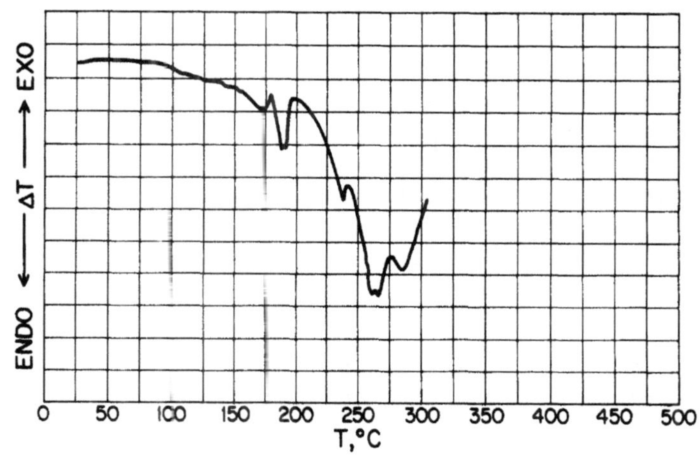

FIGURE 1f. Differential thermal analysis (DTA) hexabromocyclododecane.

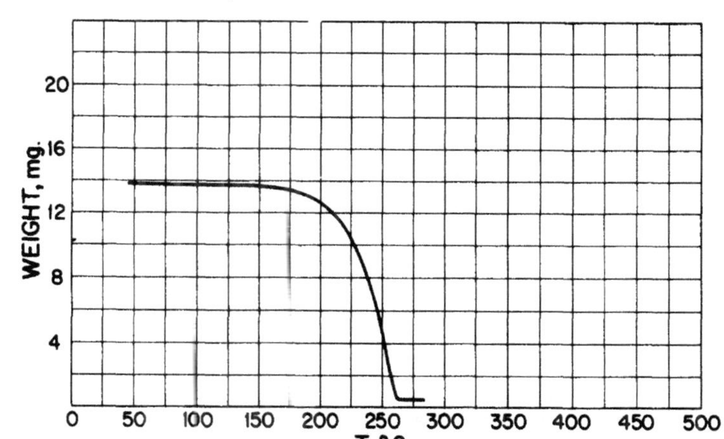

FIGURE 1g. Thermogravimetric analysis (TGA) tetrabromocyclooctane.

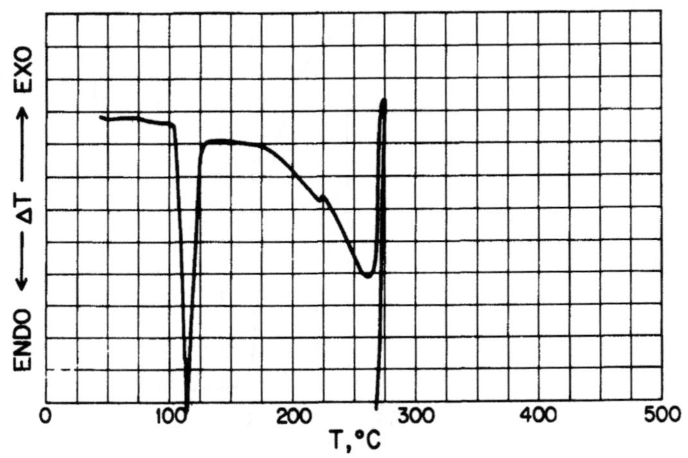

FIGURE 1h. Differential scanning calorimetry (DSC) tetrabromocyclooctane.

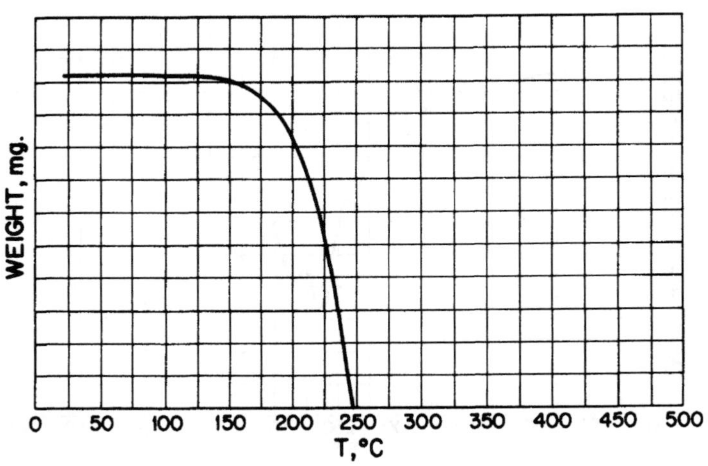

FIGURE 1i. Thermogravimetric analysis (TGA) dibromoethyldibromocylohexane.

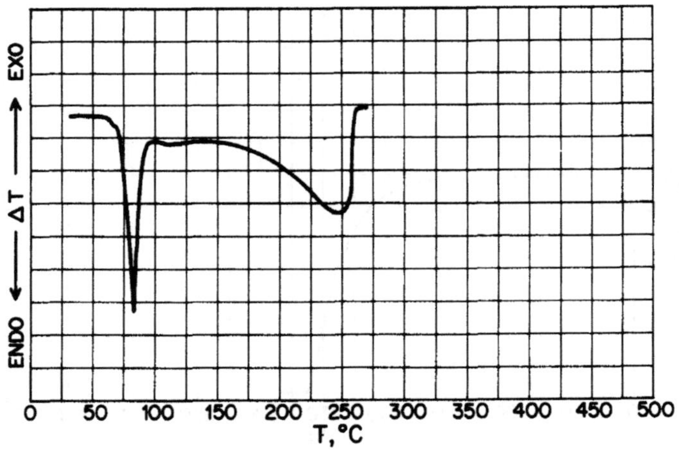

FIGURE 1j. Differential scanning calorimetry (DSC) dibromoethyldibromocyclohexane.

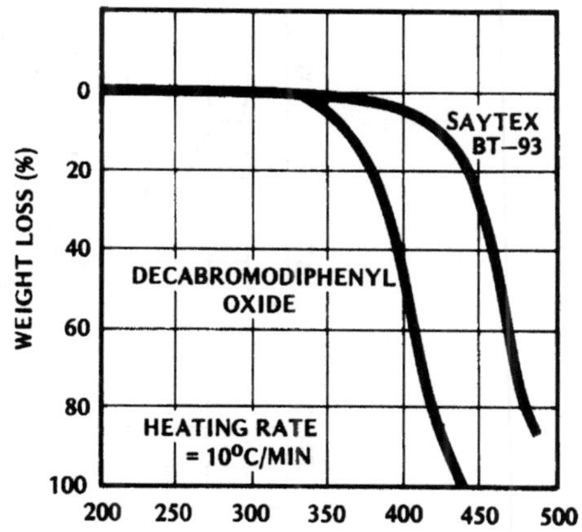

FIGURE 1k. Thermogravimetric analysis (TGA) decabromodiphenyl oxide and Saytex BT-93.

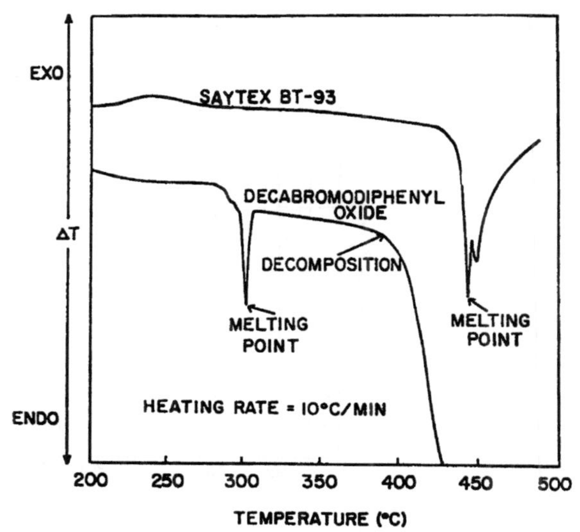

FIGURE 1l. Differential thermal analysis (DTA) decabromodiphenyl oxide and Saytex BT-93.

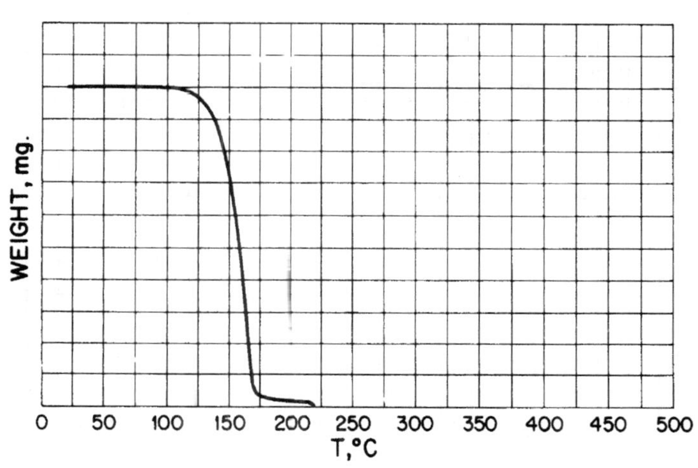

FIGURE 1m. Thermogravimetric analysis (TGA) Bromkal 73-5PE.

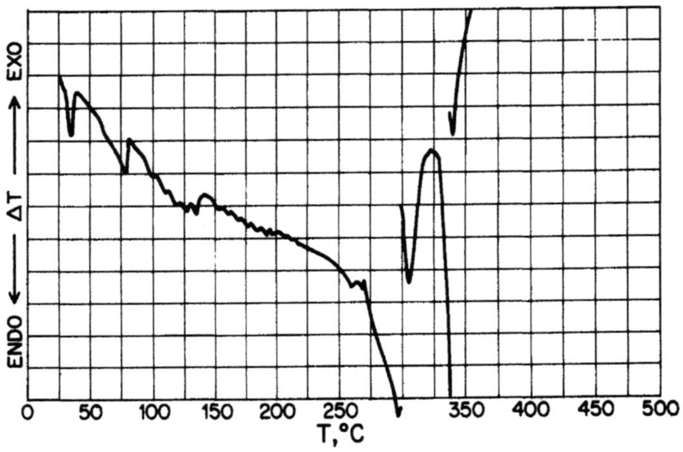

FIGURE 1n. Differential thermal analysis (DTA) Bromkal 73-5PE.

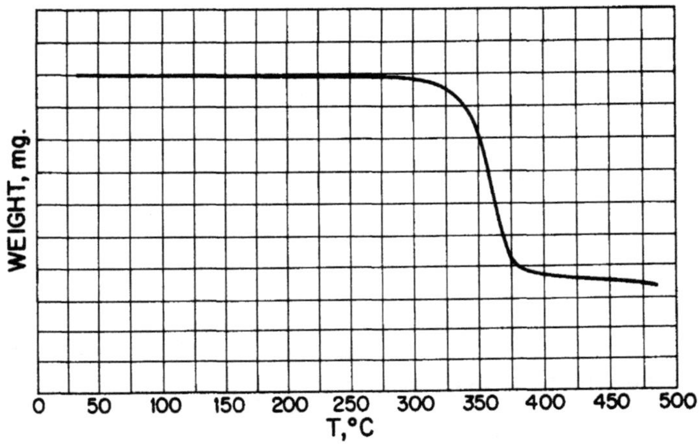

FIGURE 1o. Thermogravimetric analysis (TGA) Bromkal 66-8.

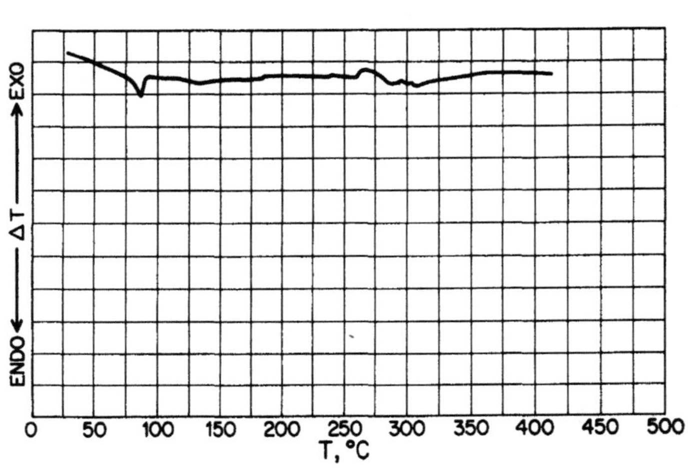

FIGURE 1p. Differential thermal analysis (DTA) Bromkal 66-8.

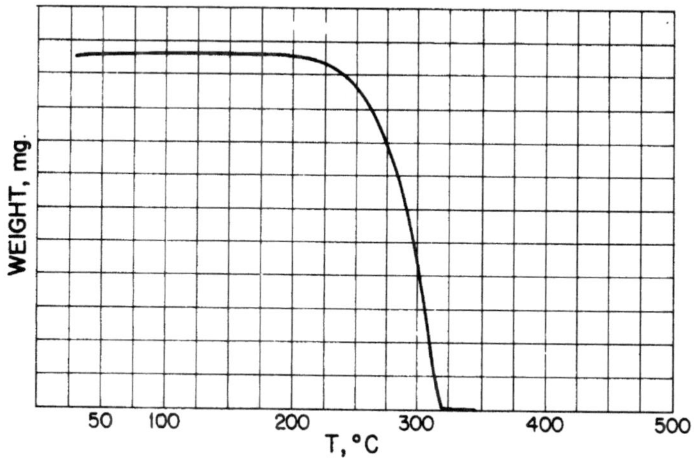

FIGURE 1q. Thermogravimetric analysis (TGA) Saytex BC-26.

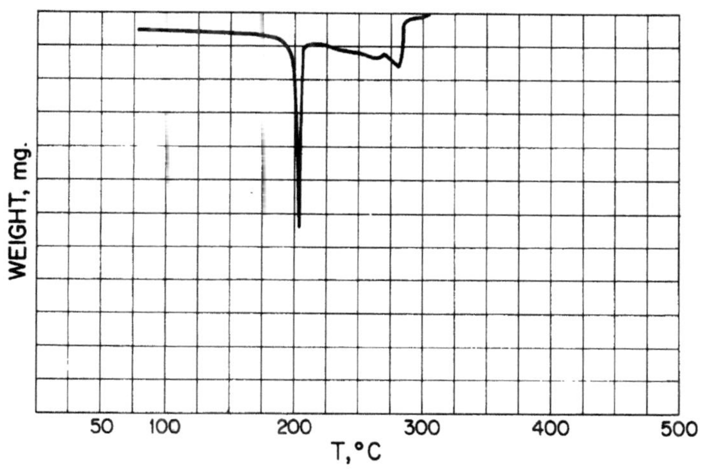

FIGURE 1r. Differential scanning calorimetry (DSC) Saytex BC-26.

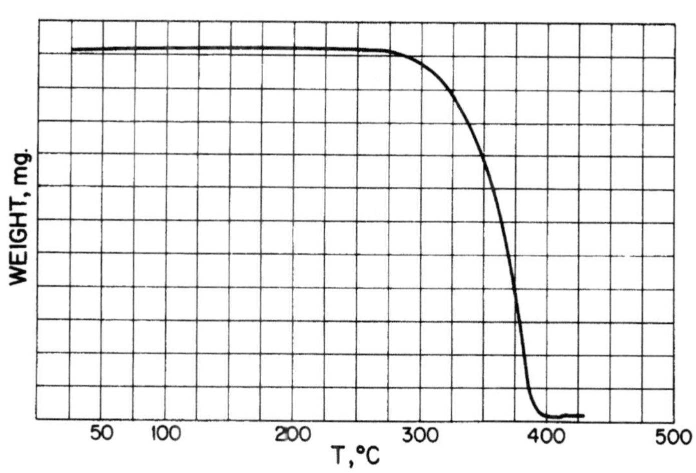

FIGURE 1s. Thermogravimetric analysis (TGA) Saytex BN-21.

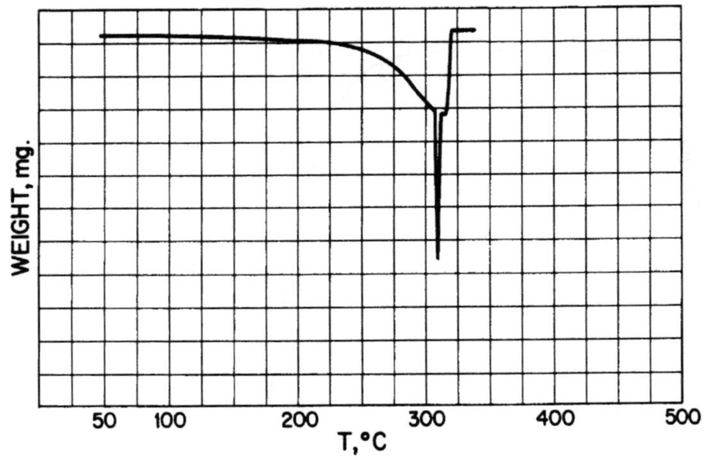

FIGURE 1t. Differential scanning calorimetry (DSC) Saytex BN-21.

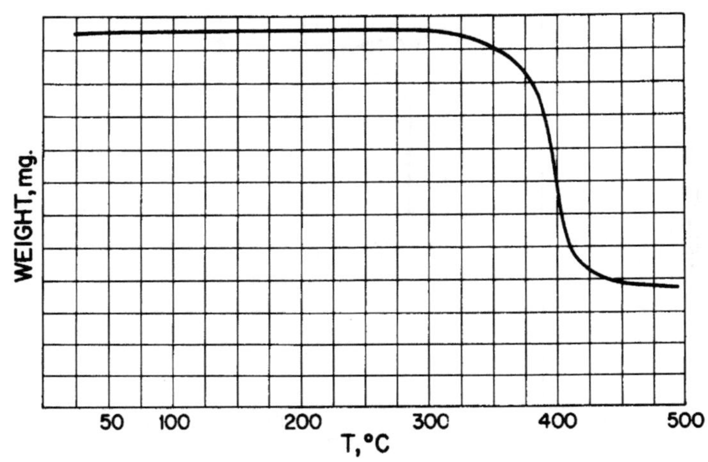

FIGURE 1u. Thermogravimetric analysis (TGA) Saytex BN-451.

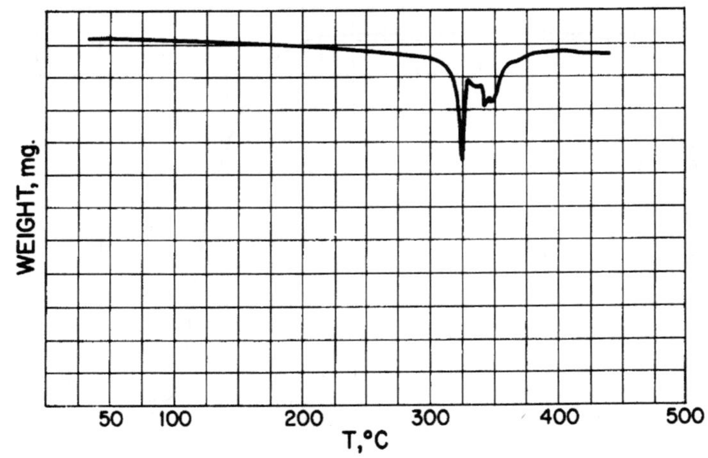

FIGURE 1v. Differential scanning calorimetry (DSC) Saytex BN-451.

and similar expressions used in this paper are not intended to reflect hazards under actual fire conditions but are only descriptive of specific laboratory tests.

In addition to flame retardancy, the effects of the additives on the properties of the polymers are discussed in detail.

2. *Flame Retardancy Tests*

The oxygen index test (ASTM D-2863) employs a vertical glass tube 60 cm high and 8.4 cm in diameter in which a rod or strip specimen is held vertically by a clamp at its bottom end.[2] A mixture of oxygen and nitrogen is metered into the bottom of the tube, passing through a bed of glass beads at the bottom to smooth the flow of gas. The sample is ignited at its upper end with a hydrogen flame, which is then withdrawn, and the sample burns like a candle from the top down. The atmosphere that permits steady burning is determined. The limiting oxygen index or simply oxygen index is the minimum fraction of oxygen in an oxygen–nitrogen mixture which will just sustain burning for 2 in. or 3 min, whichever comes first.

In the Underwriters Laboratory (UL)–Subject 94 vertical burn test a sample (5.0 × 0.05 in.) is exposed vertically to a Bunsen burner flame for 10 sec.[3] The sample is ignited at the bottom and burns up. If the specimen self-extinguishes within 30 sec, another 10-sec application is made. Flaming droplets are allowed to fall on dry absorbent surgical cotton located 12 in. below the sample. If the average burning time is less than 5 sec and the drips do not ignite the cotton, the material is classified 94V-0. If the time is less than 25 sec and the drips do not ignite the cotton, the material is classified 94V-1. If the sample is extinguishing but the cotton is ignited, the material is classified as 94V-2.

3. *Polypropylene*

One of the early applications requiring flame-retarded polypropylene was in television receiver electrical components. Underwriters Laboratories developed the UL-94 flame-retardant test, presumably to approximate the flaming conditions of an electrical short by using a small, relatively cool Bunsen burner flame for short-duration exposure. Dripping and nondripping were to be differentiated, with the latter being preferred because of the presence of other flammable components in the TV receiver and also because lint tended to build up in the cabinets. The response to this requirement was a polypropylene composition containing 27% of Dechlorane Plus and 13% antimony oxide as the synergist. Dechlorane Plus is a Diels–Alder diadduct of

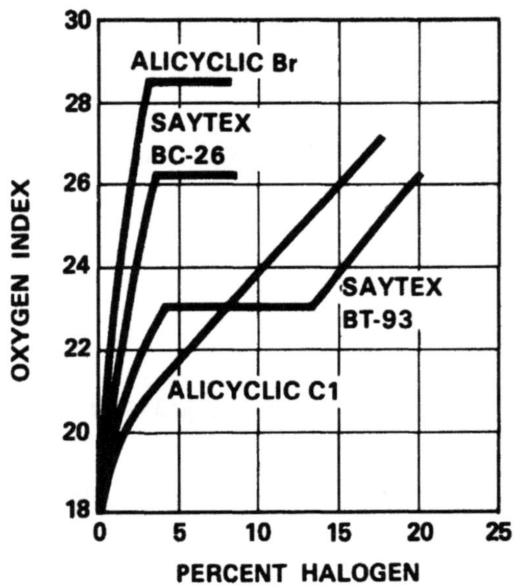

FIGURE 2. Effect of halogen in polypropylene ($FR/Sb_2O_3 = 2/1$).

2 mol of hexachlorocyclopentadiene and 1 mol of cyclooctadiene to give a compound containing 65% alicyclic chlorine. This compound has adequate thermal stability for processing into polypropylene. Figure 2 shows the effect of increasing Dechlorane Plus concentration on the oxygen index value of the plastic composition.[4] It has been found empirically that polypropylene compositions with an oxygen index of about 25.5 are necessary to obtain a self-extinguishing rating by the UL-94 test. Excessive drip or heavy charring could modify this generalization. This highly filled polypropylene composition is nondrip. Small reductions in the flame-retardant concentration result in burning and not in the anticipated self-extinguishing but dripping V-2 rating.

The composition containing 27% Dechlorane Plus and 13% antimony oxide is a nondripping composition having a UL-94 rating of V-1. This rating is due to the long glow (slow combustion) times obtained with this composition. It was later found that the afterglow could be reduced by using lower concentrations of antimony oxide and increasing the concentration of Dechlorane Plus, or replacing some of the antimony oxide with other synergists such as zinc borate. As a result, ratings of V-0 could be obtained.

A severe shortage of Dechlorane Plus occurred in 1977 and subsequently the price doubled to $1.70 per pound. A search for replacements showed the aromatic bromine-containing compounds as a viable and possible lower-priced replacement. The comparative flame retardancy of decabromodiphenyl oxide and SAYTEX BT-93 with Dechlorane Plus is shown in Fig. 2. As with Dechlorane Plus about a 40% concentration of flame retardants is required to give a nondrip UL-94 rating. Once again the interface in flame retardancy is not between V-0/V-1 and V-2, but between nondrip and burning.

Table 2 shows the comparison of three nondrip polypropylene systems.

TABLE 2
Flame-Retarded Polypropylene (V-0) Comparative FR Study

Polypropylene	51	51	51
EPDM	4	4	4
Dechlorane Plus	25	–	–
SAYTEX BT-93	–	20	–
Decabromodiphenyl oxide	–	–	20
Antimony oxide	8	6	6
Polytal 4665 (talc)	12	19	19
Oxygen index	25.8	25.8	27.5
UL-94 @ $\frac{1}{16}$ in.	V-1[a]	V-0	V-0[b]

[a]Excessive glow.
[b]Tends to drip or sag.

The standard Dechlorane Plus composition containing alicyclic chlorine as the active ingredient is compared with two aromatic bromine-containing compositions, SAYTEX BT-93 and decabromodiphenyl oxide. In an attempt to reduce the cost of this system, low-cost talc has been added. The aromatic bromine compounds will tolerate a higher concentration of talc and lower concentrations of flame-retardant additives are needed. The Dechlorane Plus compound still shows a tendency for excessive glow using sample thicknesses of 1/16 in. At these thin cross-sections, the aromatic bromine-containing compounds tend to sag with the result that the hot sagging polymer will ignite the absorbent cotton used to differentiate between dripping and nondripping. The SAYTEX BT-93 composition has less of a tendency to sag than the decabromodiphenyl oxide composition. For outdoor applications, the BT-93 compounds have better UV stability than the decabromodiphenyl oxide compounds.

In an attempt to improve the impact resistance of such highly loaded polypropylene, a low concentration of ethylene–propylene–dimer rubber was used (Table 2). This does not interfere with the flame retardancy at this low concentration. In a further attempt to improve on the impact resistance, two impact polypropylene polymers have been evaluated. The Profax 8500 series resins have high impacts due to higher levels of comonomer. As can be seen from the data in Table 3, the use of Profax 8523 (Hercules) as the base polymer results in a flame-retarded composition with an izod impact essentially equivalent to that of the virgin resin and a heat distortion temperature only slightly higher.

The literature indicates that bromine-containing compounds are far more effective flame-retardant additives than chlorine-containing compounds. As discussed above this is not so with aromatic bromine-containing compounds is polypropylene. However, as demonstrated in Fig. 2 alicyclic bromine-containing compounds are substantially more effective. A composition containing 3% alicyclic bromine will yield a polypropylene compound

TABLE 3
Talc-Filled Polypropylene Copolymer Containing SAYTEX BT-93: Physical Properties

Polypropylene homopolymer	100	–	–	–
ProFax 7523		58	–	60
ProFax 8523		–	58	–
SAYTEX BT-93		20	20	–
Decabromodiphenyl oxide				22
Antimony oxide		4	4	6
Mistron vapor (talc)		18	18	12
Oxygen index		24.8	25.2	27.8
UL-94 @ $\frac{1}{16}$ in.[a]		V-0	V-0	V-0
Izod (ft lb/in. of notch)	0.79	0.62	0.73	–
Annealed HDT (°C)	71.5	85.5	76.0	–

[a]At $\frac{1}{8}$-in. thickness, samples are V-0; at $\frac{1}{16}$-in. thickness, samples tend to sag.

with an oxygen index of about 28 and a UL-94 rating of V-2. This can be met by using SAYTEX BN-451, SAYTEX BN-21, Bromkal 66-8, and Bromkal 73-5PE. SAYTEX BC-26, a mixed bromine-chlorine compound, is also seen to be highly effective in flame-retarding polypropylene. A bromine–chlorine synergism has been suggested but this has not been demonstrated with SAYTEX BC-26.

For many years the preference has been for a nondrip polypropylene composition. This is attainable only by using about 40% flame-retardant additives. These highly loaded compositions do not possess the resiliency or properties normally associated with polypropylene. A comparison of the properties of the nondrip and drip compositions is shown in Table 4. As one

TABLE 4
Properties of Flame-Retarded Polypropylene

Polypropylene	100	94	60	56
Antimony oxide	–	2	13	6
Dechlorane Plus	–	–	27	–
SAYTEX BN-451	–	4	–	–
SAYTEX BT-93	–	–	–	20
Mistron vapor (talc)	–	–	–	18
UL-94	Burn	V-2	V-1/V-0	V-0
Izod (notched)	0.7	0.7	0.5	0.55
Heat distortion temperature				
at 66 psi, °F	226	226	245	–
at 264 psi, °F	160	–	–	209
Specific gravity	0.905	0.934	1.19	1.30
Surface	High gloss	High gloss	Dull	Dull
Tensile strength, yield, psi	4600	4400	3600	–
Flexural modulus × 10^3, psi	180	170	240	–

TABLE 5
Costs to Flame-Retard Polypropylene

Polypropylene ($0.28/lb)	94	55	56	56
Antimony oxide ($1.65/lb)	2	8	6	6
SAYTEX BN-451 ($2.70/lb)	4	–	–	–
Dechlorane Plus ($1.70/lb)	–	25	–	–
SAYTEX BT-93 ($1.60/lb)	–	–	20	–
Decabromodiphenyl oxide ($1.25/lb)	–	–	–	20[a]
Talc ($0.08/lb)	–	12	18	18
UL-94	V-2	V-1/V-0	V-0	V-0
RMC, ¢/lb	40.50	72.1	62.02	52.0
$/ft^3	23.67	55.85	50.26	42.41

[a]Poor UV resistance.

might anticipate, the V-2 composition with very low loadings of flame retardants shows little change in physical properties from those of the virgin resin. These products have a high gloss and still demonstrate the high resiliency normally associated with the virgin resin. The hinge properties also are retained. By comparison, the highly filled systems have dull surfaces, reduced impact, elevated heat distortion temperatures, and elevated flexural moduli.

Table 5 compares the cost of V-2 and V-0 polypropylene compounds both on a weight and a volume basis. As can be seen, the raw material cost for a V-0 system is substantially higher than that of a V-2 system. In addition, it would cost more money to process the V-0 system because of the higher loading of filler.

Many polypropylene applications, such as fiber, filament, and film, cannot tolerate the highly filled UL-94 V-0 type of composition. The filled compositions cannot be processed satisfactorily in the required configuration or the resultant degradation of the properties are not acceptable. As a result, UL-94 V-2 compositions are acceptable. In many areas such as polypropylene pipe, V-2 compositions have been used commercially for many years. Perhaps the most critical area is in electrical components. Even here UL-94 V-2 has become acceptable. Underwriters Laboratory recognizes that V-2 is suitable for many electrical applications in contact with live parts, provided the resistance to ignition is slightly greater than required for the nondrip system. This is illustrated in Table 6, which comes from UL Specification 1410, Television Receivers and Video Products.[5] V-2 compositions are also given yellow card listings by Underwriters Laboratory.[6]

In burning polyolefins, melt flow carries away a significant portion of the thermal energy from the flame zone. This would explain why less flame retardant is needed for a V-2 composition as compared to a nondrip composition. Melt flow not only carries away thermal energy but also

TABLE 6
Requirements for Materials in Contact with Live Parts[a]

Application	Properties					
	Resistance to Ignition from:					
	Hot bar s		Hot wire s	High-ampere arc	Burning test classification	Comparative tracking index
Available power greater than 15 W						
(A) Parts involving more than 42.4-V peak but less than 2500-V peak	10	or	10	30	94V-0	100
	15	or	15	30	94V-1	175
	30	or	30	30	94V-2	250
(B) Parts involving potentials equal to or less than 42.4 V-peak	10	or	10	–	94V-0	100
	15	or	15	–	94V-1	100
	30	or	30	–	94V-2	175
Available power equal to or less than 15 W						
(C) Parts involving less than 2500-V peak	10	or	10	–	94V-0	–
	15	or	15	–	94V-1	–
	30	or	30	–	94V-2	–
Available power–all						
(D) Parts involving potentials equal to or more than 2500-V peak	10	or	10		94V-0	–
	15	or	15		94V-1	–
	30	or	30		94V-2	–

[a]Standard for Safety, Television Receivers and Video Products, UL 1410, Underwriters Laboratories, Inc., Melville, New York (1976).

prevents the molten polymer from remaining in the burning zone preventing significant pyrolysis and preventing heat-up of the polymer adjacent to the flaming zone. One can conceive, therefore, that a V-2 system may be safer than a V-0 system in an actual fire situation. As an example, drip in a DW & V pipe system may prevent fire spreading along the pipe. Admittedly, the burning drip may become an ignition source, but laboratory experience has shown that although burning drip from a flame retarded polymer would ignite absorbent cotton 12 in. from the flame zone, it does not ignite many other types of materials. A free fall of 12 in. specified in the UL-94 test is perhaps, significant in a TV enclosure but may not be significant in other applications. It has been noted that flaming drip *containing flame retardant* will extinguish in long free falls. One can, therefore, conceive of a system design where dripping is preferred and, thereby, reduce hazard by reducing flame spread and reducing smoke.

One can conceive of a mechanism that would increase the melt flow rate of a polyolefin at high pyrolysis temperatures, thus carrying away a significant portion of the thermal energy. Such products could pass some flammability tests. This mechanism has been demonstrated with flame-retarded polystyrene by using a free-radical-generating compound, such as a peroxide or azo compound.[7] The formation of radicals on heating decomposes the polymer, thereby increasing dripping and suggesting the use of lower concentrations of flame-retardant additive.

DiPietro and Stepniczka have reported that the oxygen index of polymers decreases as the sample temperature increases.[8] Increasing the temperature of polypropylene to about 100°C can lower its oxygen index substantially. This finding is significant since in actual fire, combustible polymers are subjected to radiant heat and may be more flammable than otherwise expected.

Flame-retardant compounds and antimony oxide are very expensive and when high concentrations such as 40% are required, the result is a very expensive product. An attempt was made, therefore, using alicyclic bromine and antimony oxide at low concentration levels and a low-cost filler such as talc to reach a loading of 40%. The data (Fig. 3) show that the filled composition increases in oxygen index to a value of only about 23. This composition burns at concentrations of up to about 9% alicyclic bromine. An explanation for this is that the molten polymer does not flow away from the intense heat flux of the flame front and thus increases in temperature promoting continual pyrolysis. The flame-retardant loading is now insufficient to flame-retard the system. This would also account for the interface in the 40% loaded system being between nondrip and burning.

Antimony oxide is a required synergist for halogen-containing compounds in polypropylene. For example, a 10% concentration of aliphatic bromine with no antimony oxide gives a product with an oxygen index of 23.5

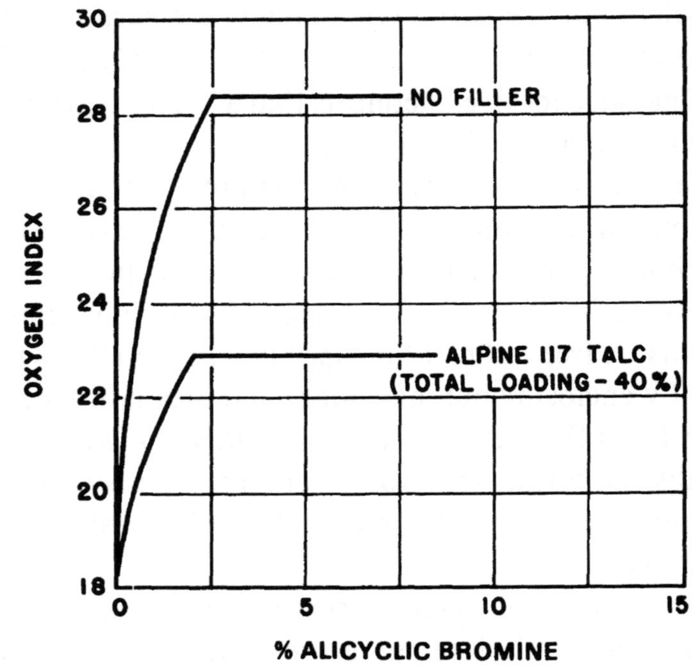

FIGURE 3. Flame-retarded polypropylene filled with talc ($FR/Sb_2O_3 = 2/1$).

(Fig. 4); this product burns in the UL-94 test. Halogen compounds with high thermal stabilities are required, to survive the normal processing temperatures of polyolefins. Such thermally stable compounds are poor flame retardants by themselves because of their low decomposition rate at the pyrolysis temperature of the polyolefin. The function of the synergist is to form volatile halides at appreciable rates at the pyrolysis temperature but to remain stable at

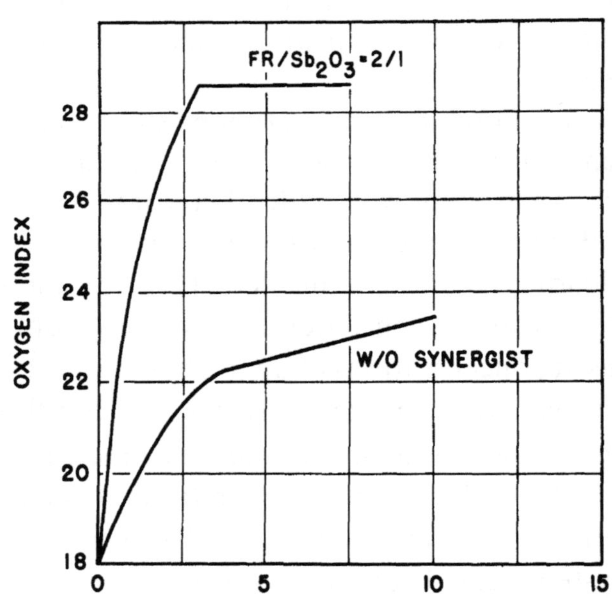

FIGURE 4. Antimony oxide requirement in polypropylene.

processing temperature. The synergism of antimony as antimony oxide with organic halogen has been well documented. Pitts *et al.*[9] have proposed that the organic halogen compound and the antimony compound react in the molten polymer to form volatile halides or oxyhalides of antimony which subsequently decompose to the halides. Because the antimony compound must react with the halogen-bearing organic compounds, dispersion of these flame retardants within the polymer has been considered to be critical. Our findings, however, have shown that this is not necessarily the case and that poor dispersions still give good flame retardancy.

It is believed that the formation of volatile antimony halides is required for synergism. Other antimony compounds are as effective as antimony oxide. The pentoxide, however, does not show the same effectivity as other antimony compounds. The general belief is that the halogen-to-antimony ratio should be about 2 for maximum effectiveness. This is only a generalization and the actual ratio for a specific system should be determined. Bismuth and arsenic trioxides also show synergism with halogen compounds. Other possible synergists include iron, boron, and tin compounds. Antimony oxide will probably remain the preferred synergist because it contains a high level of antimony and because it is essentially inert in the plastic composition during processing.

If the mechanism of formation of antimony halide or oxyhalide is interfered with then the synergist will no longer function. This is illustrated in Table 7, which shows the effect of low concentrations of silica. When 5% Cab-O-Sil is added to the polypropylene composition with an oxygen index of 26.5 and a UL-94 V-2 rating, the resultant product shows a reduction in oxygen index to 21.5 and is burning. Many color pigments have also been shown to interfere with the flame retardancy. For example, chrome pigments such as chrome yellow will interfere, presumably through the formation of stable (at pyrolysis temperature) chrome halides which render the halogen unavailable for reaction. Amine structures can also lead to the formation of ammonium halides rendering the halogen unavailable because of the relatively high decomposition temperatures. Consequently, consideration must be given to the type of pigment and other additives used in flame-retardant formulations.

Pigment additives can be difficult to disperse in polypropylene because of

TABLE 7
Effect of Silica on Oxygen Index

Polypropylene	92.5	87.5	68
Hexabromocyclododecane	5.0	5.0	18
Antimony oxide	2.5	2.5	9
Cab-O-Sil	–	5.0	5
Oxygen index	26.5	21.5	26.5
UL-94	V-2	B	V-2

the low viscosity of the polymer melt and, thereby, the absence of shear or working. Relatively poor dispersions are not objectionable from a flammability standpoint since the optimum flame retardancy is obtained. However, in the preparation of extruded pipe or polypropylene fiber good dispersion is mandatory. It is preferable that these products be first prepared as a master batch or concentrate containing approximately 50–60% of flame-retardant loading and subsequently let down into virgin polypropylene. It is preferable to prepare these concentrates using a polymer system possessing good shear during the processing. A 50/50 blend of polypropylene and low-density polyethylene or a blend of high-density polyethylene and ethylene–propylene rubber has been found satisfactory. When let down, the concentration of these polymers is very low in the final product.

4. *Polypropylene Copolymers*

It has been noted that it is much more difficult to flame retard polypropylene copolymers or blends of polypropylene homopolymer and polyethylene or rubber. It has been reported that approximately 1.5 times the flame retardant is required to flame retard copolymers. This is believed to be due to the higher pyrolysis temperature of the copolymer.[10] Figure 5 shows the effect of increasing concentration of low-density polyethylene in polypropylene on the oxygen index. As can be seen, a concentration of about 5% low-density polyethylene results in a substantial reduction in the oxygen

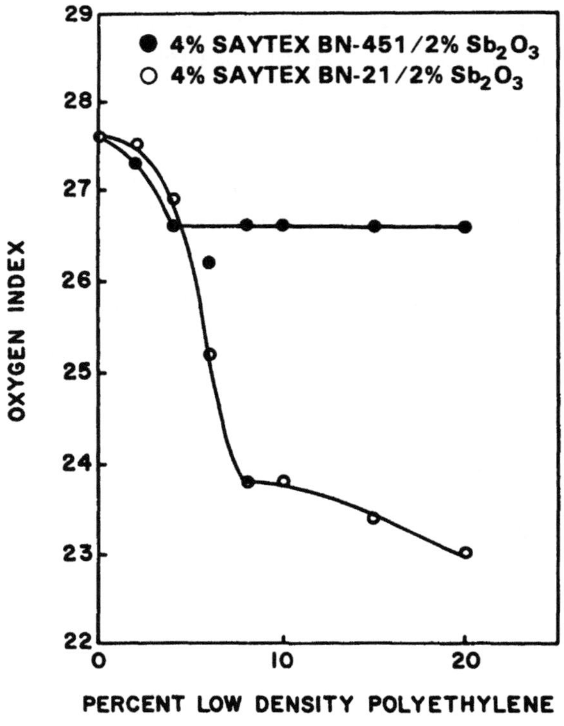

FIGURE 5. Flame-retarding polypropylene/polyethylene blends.

index. Two flame retardants are compared, SAYTEX BN-451, containing 45% alicyclic bromine, and SAYTEX BN-21, containing 40% alicyclic bromine. SAYTEX BN-451 compositions containing greater than 5% polyethylene burn and show low oxygen index values. Although the SAYTEX BN-21 system also shows a reduction in oxygen index with increasing concentration of polyethylene, the curve plateaus at an oxygen index of about 26.5 and the polymer still remains self-extinguishing by the UL-94 test. The explanation for this difference is difficult but must be related to the rate of degradation of the flame-retardant compounds. Comparison of TGA and DSC curves for these additives offers no explanation.

5. *Low-Density Polyethylene (LDPE)*

The effect of different types of halogen on flame-retarding low-density polyethylene is shown in Fig. 6. Based on oxygen index data, alicyclic bromine-containing compounds are only half as effective in polyethylene as they are in polypropylene. In contrast, aromatic bromine-containing compounds are substantially more effective in polyethylene and are perhaps the preferred flame-retardant additive. The higher pyrolysis temperature (600+°C) for polyethylene compared to polypropylene (480°C) [10] is a possible explanation for this difference. The lower thermal stability of the aliphatic bromine-containing compounds results in some loss of the active species at temperatures below the polyethylene pyrolysis temperature. The higher thermal stability of the aromatic bromine-containing compounds is more consistent with the temperature of the pyrolyzing polyethylene. The

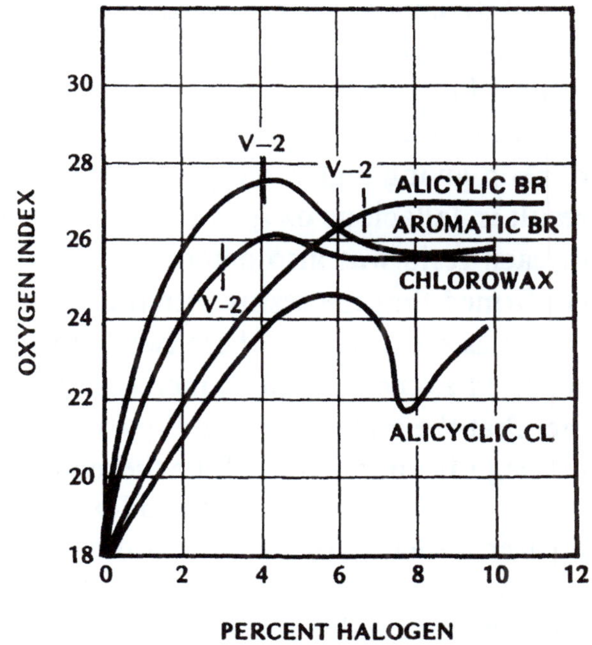

FIGURE 6. Effect of halogen type on LDPE (MI-4.5) flammability (FR/Sb_2O_3 = 2/1).

high stability of the aromatic bromine-containing compounds apparently does not allow for maximum concentration of the active species at the low pyrolysis temperature of polypropylene. This concept does not explain the high efficiency of the chlorinated paraffins. Compositions containing 3% chlorinated paraffin have an oxygen index of about 25 and a UL-94 rating of V-2.

The alicyclic chlorine-containing compound Dechlorane Plus shows maximum and minimum peaks with the oxygen index increasing again as the flame-retardant concentration is increased further to high levels. At low concentrations of additive much thermal energy is presumably carried away from the flame zone by melt flow. The efficiency of Dechlorane Plus is poor and higher concentrations are required. As the additive level is increased, thickening of the polymer melt reduces the melt flow and the oxygen index drops. With further increase in additive concentration the flammability is overcome and the oxygen index rises once more.

The poor thermal stability of chlorinated paraffin (about 190°C) restricts the use of this additive essentially to blown LDPE film. Another problem attributable to the poor thermal stability of the additive is the difficulty in reusing the scrap which discolors upon reprocessing. However, the very low cost of the chlorinated paraffins and its high efficiency make it a favorite flame-retardant additive for polyethylene blown film. In injection molding, extrusion coating, and wire and cable applications, Chlorowax cannot be used and either Dechlorane Plus or aromatic bromine-containing compounds are required. For extrusion coating applications, where processing temperatures reach about 300°C, the only suitable additive is SAYTEX BT-93. This additive is stable to temperatures of about 450°C and disperses quite well in the polyethylene.

Attempts to flame retard different low-density polyethylene resins have given a considerable variability in results. This can be seen in Fig. 7, in which resins of different melt indices were used. The oxygen index values appear to be a function of the melt index of the resin with higher melt index or flow rate of resin giving higher oxygen index values. The consensus that drippy resins result in fallacious oxygen index values was considered and the flammability of each resin was also determined by the UL-94 vertical burn test. This test confirmed that the resins with the higher melt index would be easier to flame retard as measured by these two laboratory tests, and less flame retardant is required as the melt index increases. Once again, this can be explained by the fact that the higher the melt flow of the polymer the greater would be the loss of a significant portion of the thermal energy from the flame zone. The higher crystallinity of the low-melt-index polyethylene may also contribute to the greater difficulty of flame retarding them.

A closer comparison of the flammability of two low-density resins of differing melt index is shown in Fig. 8. The flame-retardant additive contains

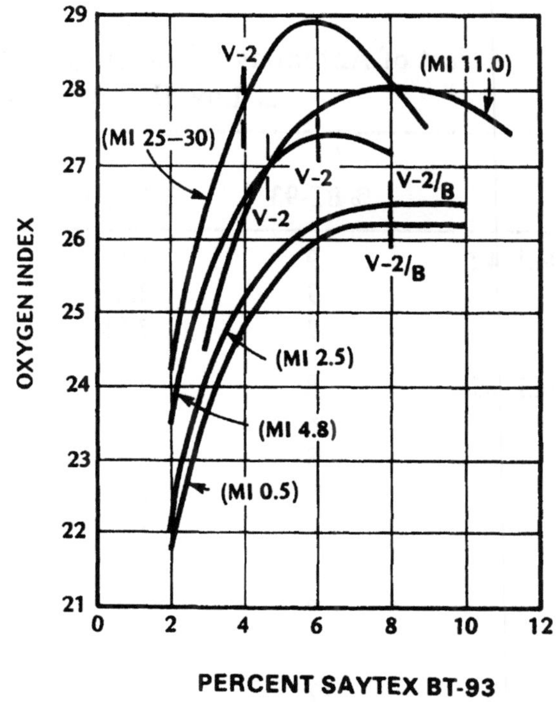

FIGURE 7. Effect of Saytex BT-93 on oxygen index of low-density polyethylene (FR/Sb_2O_3 = 2/1).

aromatic bromine. As can be seen, higher oxygen index values can be obtained using the higher melt flow resin. Evaluation in the UL-94 test shows that about 4.5% SAYTEX BT-93 is required to give a V-2 rating for the 11 melt index resin, whereas 6% additive is required for the 0.5 melt index resin. The oxygen index values for the two resins at these V-2 points differ by about 1 unit. From a practical application, however, one may wish to use the same concentration of flame retardant for each resin. The UL-94 V-2 points fall upon the steep slope of the curves. In production, a slight error in weighing or poor dispersion could result in a burning composition. Operation at the peak

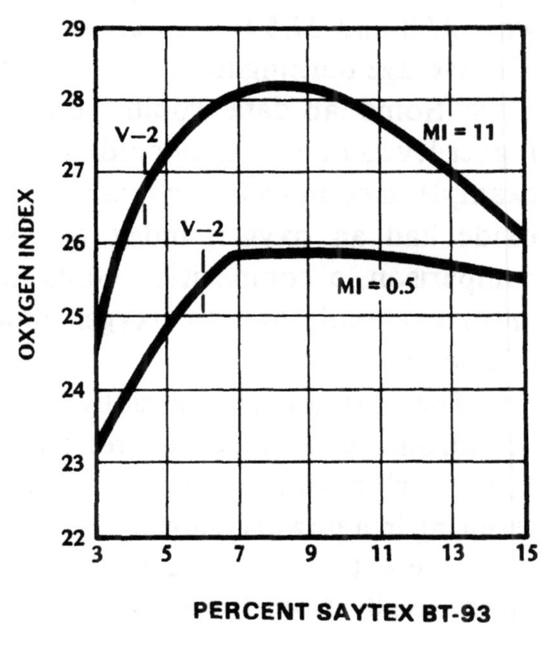

FIGURE 8. Effect of aromatic bromine on the oxygen index of LDPE (FR/Sb_2O_3 = 3/1).

TABLE 8
Effect of Antimony Oxide Concentration on Flame-Retardant Properties of Low-Density Polyethylene (MI = 4.5)

	% BT-93	% Sb_2O_3	Avg. burn time in UL-94 (sec)	UL-94	Oxygen index
M.I. 4.5	9	4.5	3	V-2	26.2
	9	3.0	1	V-2	26.9
	9	2.25	<1	V-2	27.2
M.I. 25	8	4	1	V-2	27.8
	8	3	<1	V-2	27.8
	8	2	1	V-2	27.5
	8	1	17	V-B	27.2
	3	3	>25	V-B	24.5
	3	2	14	V-B	25.2
	3	1.5	5	V-2	25.8
	3	1	16	V-B	24.5
M.I. 0.5	12	6	–	V-B	25.2
	6	3	–	V-B	25.8
	6	2	–	V-2	25.5

or plateau of the curve with both resins is consistent with good practice, and the use of approximately 7% flame-retardant additive is recommended.

The effect of antimony oxide concentration on the flame retardancy of LDPE is shown in Table 8. These data show that antimony is much more efficient in polyethylene than we had seen earlier in polypropylene. A 4:1 ratio of FR–antimony oxide is quite sufficient to give a UL-94 rating of V-2 with an oxygen index of 27.2, which is as good if not better than the compound containing a 2:1 ratio. Using the 25 melt index resin the oxygen index values were 27.8 and 27.5 for 2:1 and 4:1 FR–antimony oxide ratios, respectively. The average burning time in the UL-94 test for both of these samples was only 1 sec. Some lab data would indicate that with a 0.5 melt index LDPE the higher levels of antimony oxide were actually detrimental to the system. For example, a composition containing 12% SAYTEX BT-93 and 6% antimony oxide had an oxygen index of 25.2 but burned in the UL-94 test. In comparison, a composition containing only 6% SAYTEX BT-93 and 2% antimony oxide had an oxygen index of 25.5, with UL-94 rating of V-2 (Table 8).

The brominated cyclohydrocarbons (alicyclic bromine) such as hexabromocyclododecane, tetrabromocyclooctane, and dibromoethyldibromocyclohexane show substantial bloom from low-density polyethylene resulting in a powdery surface. Aesthetically this is undesirable but the bloom also presents some practical problems. Flame retardant can be lost mechanically from the surface reducing flame retardancy. Also the bloom tends to pick up lint and moisture affecting electrical properties. Decabromodiphenyl oxide also shows substantial bloom from polyethylene. The bloom with this high-melting pigment was not anticipated; generally soluble, low-melting additives bloom from many polymers. Although Chlorowax systems show a high gloss, the polyethylene film has an oily feel resulting in some problems in heat sealing and surface printing. The high crystallinity of polyethylene is apparently responsible for this blooming activity as compared to the lesser bloom in polypropylene. SAYTEX BT-93 shows no bloom from polyethylene and has been used satisfactorily even in extrusion coating applications.

As with polypropylene, seemingly inert additives can interfere with the flame retardancy of the plastic. The addition of 5% silica has a dramatic effect on the oxygen index of a high-density polyethylene system (Fig. 9). The effect of the silica can be overcome by a substantial increase in the antimony oxide concentration in the system.

Extrusion coated products are processed at a temperature of about 300°C and require good adhesion to the substrate. SAYTEX BT-93 has been successful in this application because of its high thermal stability.

Thin films burn differently than thicker configurations. Film or fiber shrink away from the flame and ball up, the ignition finally taking place on the thicker portion. Oxygen index values of thin films tend to be higher because of this shrinking effect.

Attempts to form a low-cost nondrip UL-94 V-0 polypropylene by the addition of low-cost inert fillers such as talc were unsuccessful. With low-density polyethylene, however, this method has been successful. As can be seen in Table 9, the use of 12% Saytex BT-93, 6% antimony oxide, and 30% inert talc gives a product with an oxygen index of 28.1 and a UL-94 rating of

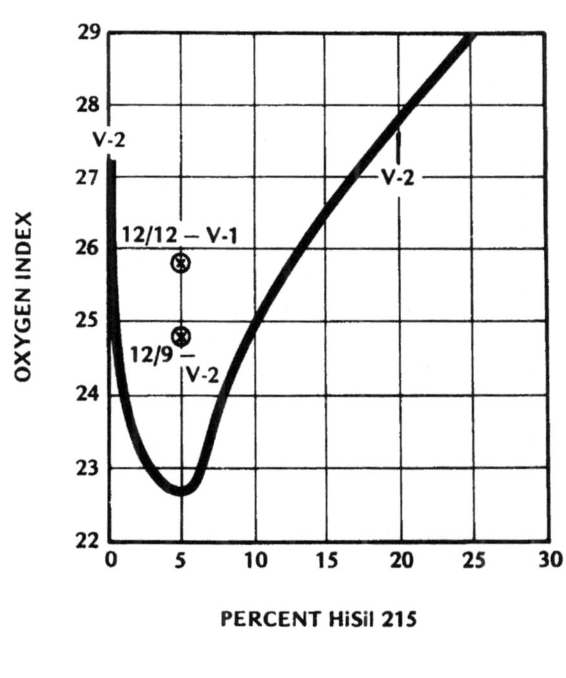

FIGURE 9. Effect of silica filler on flammability of high-density polyethylene containing 12% Saytex BT-93.

TABLE 9
Low-Density Polyethylene
(UL-94 V-0)

LDPE	52
SAYTEX BT-93	12
Antimony oxide	6
Talc	30
Oxygen index	28.1
UL-94	V-0
first Ign. (sec)	0-0
second Ign. (sec)	1-3

V-0. The reduction in the flame retardant's effectiveness in a filled system apparently is not quite as severe in polyethylene as it is in polypropylene.

6. *High-Density Polyethylene*

The effect of various types of halogen compounds in high-density polyethylene is shown in Fig. 10. These data are consistent with what had been shown above with LDPE (Fig. 6). Once again, the alicyclic compound, specifically, Dechlorane Plus, shows a maximum and minimum. Also, Chlorowax seems to be a very effective flame-retardant additive. Its application in high-density polyethylene is severely limited since most applications require high-temperature processing. Once again, it would appear that aromatic bromine-containing compounds are the additives of choice. The effect of melt index and density differences were evaluated and the

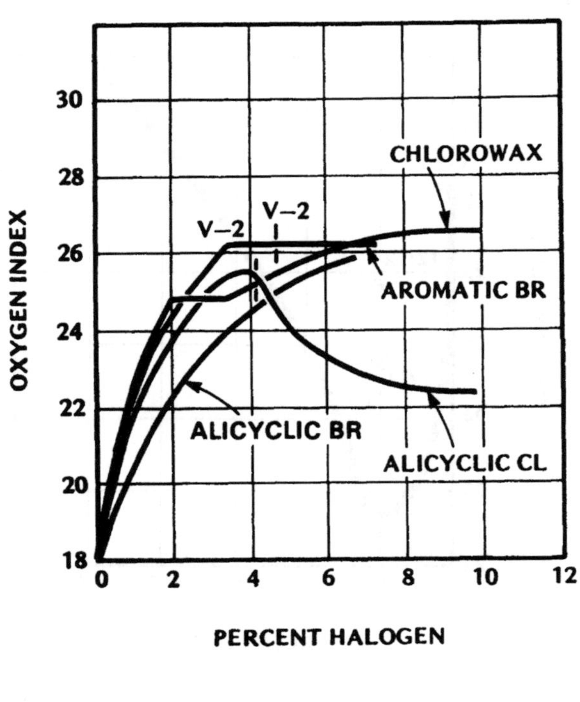

FIGURE 10. Effect of halogen type on high-density polyethylene flammability ($FR/Sb_2O_3 = 2/1$).

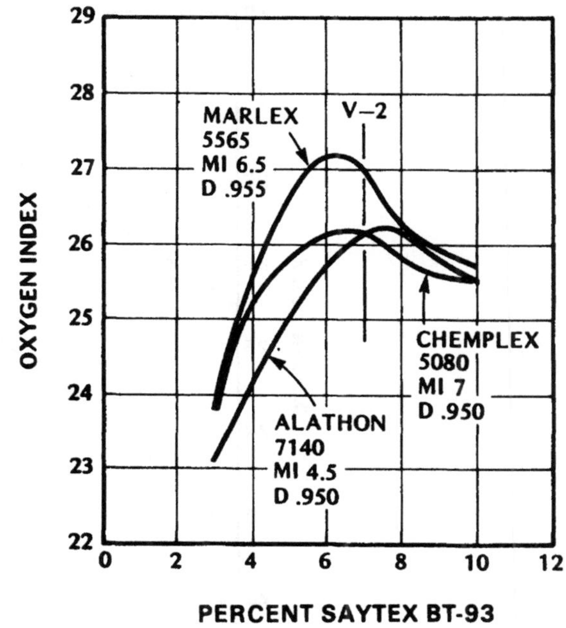

FIGURE 11. Effect of Saytex BT-93 concentration on oxygen index of various high-density polyethylene resins (FR/ Sb_2O_3 = 2/1).

data shown in Fig. 11. Here, factors other than melt index seem to be responsible for oxygen indices that vary considerably. However, with all three resins 6% Saytex BT-93 is required to attain a UL-94 V-2 rating. Figure 12 compares the effectiveness of antimony oxide in high-density polyethylene. From an oxygen index standpoint a 4:1 ratio appears to be the most effective; however, the UL-94 test would indicate that a 3:1 ratio is optimum. As with low-density polyethylene the higher pyrolysis temperature apparently promotes a greater synergism between antimony and stable halogen compounds, resulting in even more efficient use of antimony and a lower antimony requirement. Figure 13 shows that a V-0 high-density polyethylene is easily

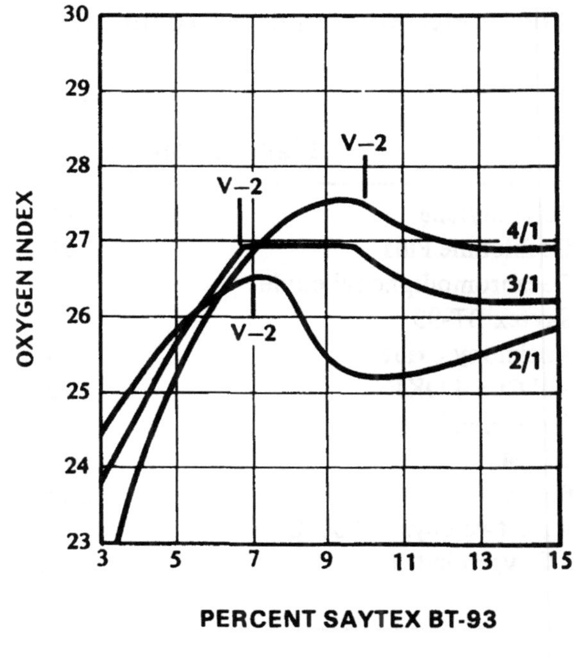

FIGURE 12. Effect of FR/Sb_2O_3 on oxygen index of high-density polyethylene containing Saytex BT-93.

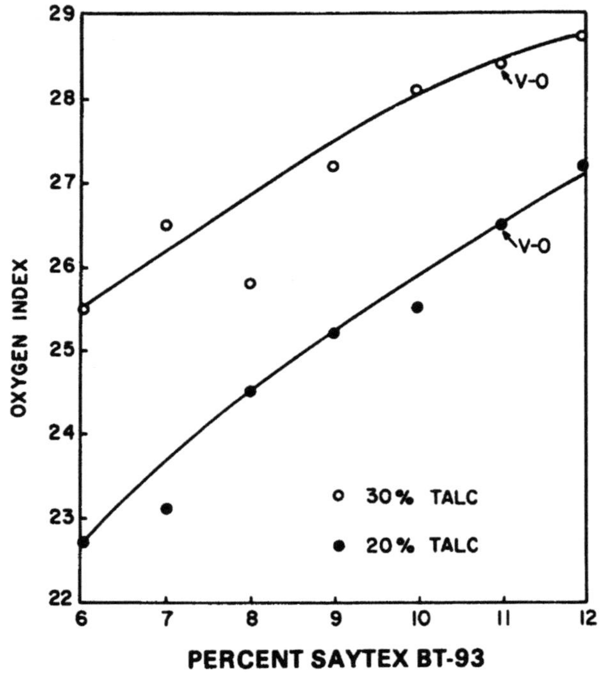

FIGURE 13. Effect of Saytex BT-93 concentration on the oxygen index of Carbide 5565 high-density polyethylene containing polytal talc ($FR/Sb_2O_3 = 2/1$).

attainable by the addition of large quantities of low-cost inert talc. As can be seen, increased levels of flame retardant and antimony oxide are required in comparison to the V-2 systems, but the overall composition is low in cost.

7. *Cross-Linked Polyethylene*

Polyethylene can be cross-linked with radiation or with peroxides such as dicumyl peroxide. One of the major applications is wire and cable insulation. The preferred flame retardant has been the alicyclic chlorine-containing

TABLE 10
Flame-Retarded Cross-Linked Polyethylene

Polyethylene	100	100	100
Dechlorane Plus	30	–	–
Decabromodiphenyl oxide	–	30	–
Saytex BT-93	–	–	30
Antimony oxide	15	15	15
Vul-Cup 4 OKE	5	5	5
Oxygen index	24.6	26.1	26.1
UL-94	Burns	V-0	V-0
Bloom	Slight	Very heavy	None
Aged two days at 160°C			
Oxygen index	25.4	–	27.6
UL-94	Burns	V-0	V-0

TABLE 11
Effect of Sb_2O_3 Loading on Oxygen Index of Flame-Retarded Cross-Linked Polyethylene

Polyethylene	100	100	100	100	100	100
SAYTEX BT-93	30	30	30	30	30	30
Sb_2O_3	2.5	5	7.5	10	12.5	15
Vul-Cup 4 OKE	5	5	5	5	5	5
Oxygen index	22.7	23.2	23.9	24.6	24.6	24.6
UL-94	B	B	V-1	V-0	V-0	V-0

Samples are 70 mils thick cross-linked at 360°F for 20 min.

compound Dechlorane Plus. Recent requirements to meet more rigid specifications have resulted in the development of cross-linked compositions containing aromatic bromine flame retardants. Oxygen index and UL-94 measurements of slab stock show the Dechlorane Plus system to be deficient in flame-retardant properties. However, when coated wire is tested by the VW-1 flame test, the system passes.

Table 10 compares the flammability of cross-linked polyethylene compositions containing Dechlorane Plus, decabromodiphenyl oxide, and Saytex BT-93. The Dechlorane Plus composition has a low oxygen index and burns in the UL-94 test.

Most commercial grades of decabromodiphenyl oxide are contaminated with inorganic salts as a result of acid neutralization during production. They also contain relatively high levels of iron contamination. The apparent result is degradation of electrical properties during high-temperature water aging.

Electrical insulation of unreinforced hydrocarbon polymers such as cross-linked polyethylene are known to fail during hot water aging.[11-13] Under the influence of electrical stress water trees develop as water molecules push aside the polymer chains. The water agglomerates as disconnected droplets of water. Such electrical property degradation is not seen when decabromodiphenyl oxide of low salt content (commercially available as Saytex 102E) or when Saytex BT-93 is used. It is proposed that the salt contamination at least accelerates this treeing mechanism.

TABLE 12
SAYTEX BT-93 in Cross-Linked Polyethylene

LDPE	100	100
SAYTEX BT-93	35	35
Antimony oxide	17.5	17.5
MT carbon black	–	50
Vul-Cup 4 OKE	4	4
UL-94–$\frac{1}{16}$ in.	V-0	V-0
Oxygen index	28.5	29.6

TABLE 13
Effect of BT-93 Concentration on Oxygen Index of Surlyn 1650

Surlyn 1650	100	92.5	89.5	86.5
SAYTEX BT-93	–	5	7	9
Antimony oxide	–	2.5	3.5	4.5
Oxygen index	18.0	22.4	24.8	27.5
UL-94	Burn	Burn	V-2	V-2

Saytex BT-93 in cross-linked polyethylene gives a unique product with excellent flame resistance and good high-temperature water aging. Table 11 shows the optimum flame retardant to antimony oxide ratio to be 3:1. Medium thermal carbon black filler does not interfere with flame retardancy (Table 12) as do some high reinforcing blacks.

8. *Ionomers*

Ionomers are copolymers of ethylene containing carboxyl groups. These polymers are cross-linked through ionic crosslinks. The polymers are processed at conventional temperatures in standard injection molding and extrusion equipment like other thermoplastics. Ionic interchain forces give the thermoplastic polymer properties normally associated with a cross-linked structure.

Surlyn 1650, a polyethylene copolymer, can be flame retarded using 7% Saytex BT-93/3.5% antimony oxide. This composition shows an oxygen index of 24.8 and a UL-94 rating of V-2 (Table 13). A 9:4.5 composition has an oxygen index of 27.5.

9. *Polybutylene*

Polybutylene resins are high-molecular-weight isotactic polymers of butene-1. These crystalline thermoplastics offer a combination of flexibility, toughness, stress crack resistance, and creep resistance. Polybutylene resins are available in general purpose homopolymer grades and in two copolymer grades of 1 and 2 melt index. They are finding use in pipe, film, molding, and alloying applications.

Polybutylene of 2 melt index has been flame retarded using Saytex BN-451 (Fig. 14). The oxygen index–Saytex BN-451 concentration curve shows a maximum at about 5% flame retardant with oxygen index dropping off sharply at both lower and higher concentraton. The composition contains antimony oxide at a flame retardant–antimony oxide ratio of 2:1. The highly crystalline nature (50–55%) of polybutylene may be responsible for the difficulty in flame retarding this resin.

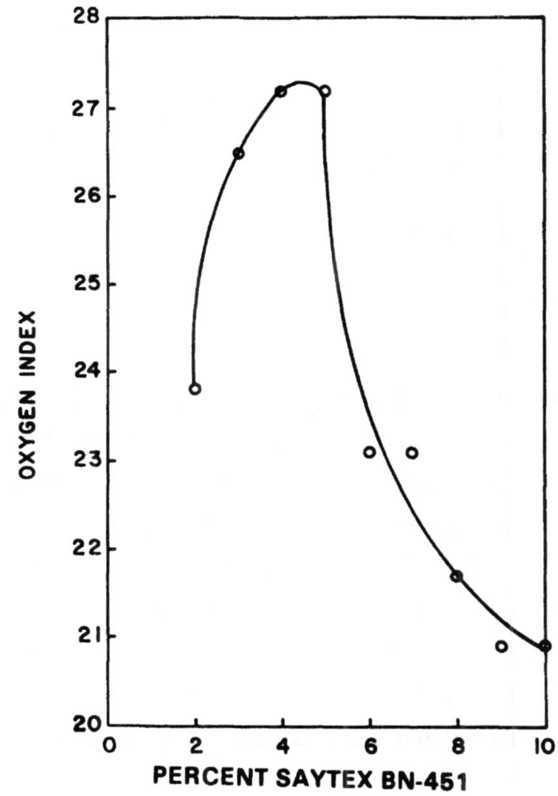

FIGURE 14. Flame-retarding polybutene (1200S).

10. *Ethylene–Propylene–Dimer Rubber*

The efficiency of three flame retardants in ethylene–propylene–dimer rubber is compared in Fig. 15. Based upon oxygen index data, alicyclic chlorine-containing compounds are inefficient (the slope of the oxygen index–concentration curve is low). Both aromatic bromine-containing compounds

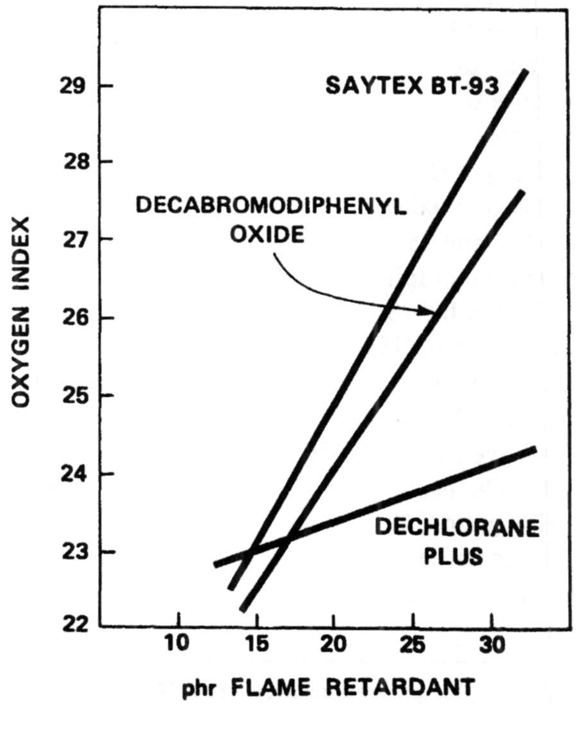

FIGURE 15. Effect of different types of halogen-containing compounds on oxygen index of EPDM rubber ($FR/Sb_2O_3 = 2/1$).

TABLE 14
Flame-Retarded EPDM

Vistalon	100	100	100	100	100
SAYTEX BT-93	30	30	30	30	30
Sb_2O_3	5	7.5	10	12.5	15
Vul-Cup 4 OKE	6	6	6	6	6
Oxygen index	24.6	25.4	27.6	27.6	28.5
UL-94	B	B	94V-0	94V-0	94V-0

show about the same efficiency, although the lower bromine content additive appears more efficient. The slopes of both curves are relatively high compared to Dechlorane Plus. A flame retardant–antimony oxide ratio of 3:1 is optimum for obtaining a UL-94 V-0 rating (Table 14).

Cured formulations for white and black EPDM compounds containing Dechlorane Plus, decabromodiphenyl oxide, and Saytex BT-93 are shown in Tables 15 and 16. The tensile data show the relatively poor properties of the Dechlorane Plus compound. The black decabromodiphenyl oxide compound shows poor heat-aging properties and bloom of the flame retardant.

11. Ethylene–Propylene–Dimer (EPDM) Rubber Wire and Cable Insulation

Many wire and cable constructions require a combination of toughness, good mechanical strength, excellent electrical stability, good flame resistance,

TABLE 15
Flame-Retarded EPDM (White Filled)

Vistalon 3708	100	100	100
Sun Par 150	15	15	15
Zinc oxide	5	5	5
Agerite D	1.5	1.5	1.5
Whitetex	110	110	110
Vul-Cup 4 OKE	6	6	6
Antimony oxide	15	15	15
Dechlorane 25	30	–	–
Decabromodiphenyl oxide	–	30	–
SAYTEX BT-93	–	–	30
Cured 30 min/166°C			
Tensile, psi	790	1220	1310
Elongation, %	570	625	620
Oven aged 70 hr/121°C			
Tensile, psi	800	1120	1250
Elongation, %	520	610	610
Oxygen index	26.1	26.1	26.1
UL-94 burn times, sec[a]	9–2	0–1	0–3
	16–4	0–1	0–3

[a]All samples glowed greater than 60 sec.

TABLE 16
Flame-Retarded EPDM (Black Filled)

Vistalon 7308	100	100	100
Sun Par 150	15	15	15
Zinc oxide	5	5	5
Agerite D	1.5	1.5	1.5
Statex M550	55	55	55
Vul-Cup 4 OKE	6	6	6
Antimony oxide	15	15	15
Dechlorane 25	30	–	–
Decabromodiphenyl oxide	–	30	–
SAYTEX BT-93	–	–	30
Cured 25 min/160°C			
Tensile, psi	1980	2220	2340
Elongation, %	370	450	410
Oven aged 70 hr/121°C			
Tensile, psi	1770	2090	2510
Elongation, %	310	340	400
Oxygen index	26.8	27.6	27.3
UL-94 burn times, sec[a]	0–0	0–0	0–0
	0–0	0–0	0–0

[a]All samples glowed greater than 60 sec.

and elastomeric flexibility before and after heat aging. Such a combination has been achieved through the use of a composite construction of a separate jacket over an insulation. This combination has performed with a highly reliable service record. However, a genuine need has developed for reduced wall constructions with improved thermal properties, especially for nuclear station cables and secondary network cable.

An ethylene–propylene–dimer rubber, Nordel 2722, has been com-

TABLE 17
Nuclear Station Cable[a] (#12AWG Aluminum with 1.19-mm Insulation)

	Composition A
EPDM	90
Low-density polyethylene	20
Zinc oxide	5
90% red lead dispersion	5
Paraffin wax	5
Zinc salt of mercaptobenzimidazole	2
Low-temperature reaction product of acetone and diphenylamine	1
Treated calcined clay	60
Vinylsilane	1
SRF black	2
Diadduct of hexachlorocyclopentadiene and cyclooctadiene (Dechlorane Plus)	33
Antimony oxide	12
Dicumyl peroxide	3

[a]Reference 13.

TABLE 18
Nuclear Station Cable[a] (#12AWG Aluminum with 1.19-mm Insulation)

Composition B	
EPDM	90
Low-density polyethylene	20
Zinc oxide	5
90% litharge dispersion	5
Paraffin wax	5
Zinc salt of mercaptobenzimidazole	2
Low-temperature reaction product of acetone and diphenylamine	1
Treated calcined clay	60
Vinylsilane	1
Antimony oxide	12
SAYTEX BT-93	32
a,a'-bis(*t*-butylperoxy)diisopropylbenzene	2.8

[a]Reference 13.

pounded to give flame-retarded EPDM integral insulation jacket compositions with excellent heat resistance, mechanical strength, and electrical stability.[14] The semicrystalline nature of this EPDM polymer with its toughness and physical strength makes possible reduced wall constructions. The use of Dechlorane Plus and SAYTEX BT-93 as flame-retardant additives give the desired combination of good physical properties, VW-1-type flame performance, stable wet electricals, heat resistance, and acceptable cost.

The performance and reliability of class 1E cable in nuclear power generating stations is of great concern today. To meet industry demands of reduced-wall, improved-performance constructions, candidates very similar to compositions A and B shown in Tables 17 and 18 have been qualified for nuclear control cable and instrumentation wire following IEEE-383-type testing. Their superior heat resistance makes it possible to predict a service life of over 40 years as a 90°C insulation as determined from Arrhenius plots of aging data. Similar compositions have been qualified by Consolidated Edison of New York as a Unipass secondary network cable. They have performed very well in the critical areas of rate of oxidation, damage-threshold-type roasting, and bonfire flame tests. The test data is given in Ref. 14.

Compositions C and D of Tables 19 and 20 have been developed to meet the multilisting of UL constructions like XHHW, RHH–RHW, USE, and SIS a-1 with VW-1 flame ratings. Once again, the flame retardants of choice include Dechlorane Plus and SAYTEX BT-93.

Properly reinforced amorphous ethylene–propylene rubber insulated cables are expected to be more stable in a wet environment than unfilled cross-linked polyethylene insulated cables. They do not fail by the formation of water trees as the primary mechanism, apparently due to the strong bonds between filler and polymer.[15] Fillers may have to be surface treated for proper reinforcement.[14,15]

TABLE 19
Type XHHW[a] (#14 AWG Solid Bare Copper with 0.76-mm Insulation)

Composition C	
EPDM	90
Low-density polyethylene	20
Zinc oxide	5
Paraffin wax	5
90% litharge dispersion	5
Oligomeric 2,2,4-trimethyl-1,2-dihydroquinoline	1.5
Calcined clay	60
Vinylsilane	1.5
Diadduct of hexachlorocyclopentadiene and cyclooctadiene (Dechlorane Plus)	33
Antimony oxide	12
2:1 Complex of alkylated phenol phosphonate with nickel	0.75
40% Dispersion in treated clay of a,a′-bis(*t*-butylperoxy)diisopropylbenzene	8.0

[a]Reference 13.

It has been proposed that the full reinforcement of rubber by carbon black requires the formation of strong chemical bonds between the carbon black particles and the polymer chain.[16] The weaker physical bonds involved in absorption of rubber molecules on carbon black play a small role in property improvement. With carbon blacks, chemical bonds involve free radicals formed from mechanical shearing forces on the polymer during processing. The reinforcing effect of nonblack materials such as clays, carbonates, and silicas can also be improved by the formation of strong chemical bonds. One method of doing this is to use certain silanes with dual functionality.[14,15] When clay is treated with trimethoxy ethoxy vinyl silane, the alkoxy groups of the silane hydrolyzes with traces of water to form silanols that react with oxy or residual hydroxy groups on the clay surface. During free

TABLE 20
Type XHHW[a] (#14 AWG Solid Bare Copper with 0.76-mm Insulation)

Composition D	
EPDM	90
Low-density polyethylene	20
Zinc oxide	5
Paraffin wax	5
90% litharge dispersion	5
Oligomeric 2,2,4-trimethyl-1,2-dihydroquinoline	1.5
Calcined clay	60
Vinylsilane	1.5
SAYTEX BT-93	32
Antimony oxide	12
2:1 Complex of alkylated phenol phosphonate with nickel	0.75
40% Dispersion of a,a′-bis(*t*-butylperoxy)diisopropylbenzene in treated clay	8

[a]Reference 13.

radical cross-linking of the EP rubber, some free radical polymer chains react with the free radicals formed from vinyl groups of the silane.

Ethylene–propylene–dimer rubber filled with silica or clay under conditions such that no chemical bond could be formed gives a product with poor electrical stability. For optimum electrical performance, it is essential that either treated calcined clay or calcined clay in combination with reactive-type silane coupling agent such as a vinylsilane-type coupling agent, be used. This chemical bonding of inorganic filler–organic polymer presumably makes a composition with a continuous organic phase which reduces its affinity toward water. Calcined clay, in combination with vinylsilane, is adequate for electrical properties in 75°C water. However, for long-term electrical stability, especially in 90°C water, treated calcined clays in combination with vinyl silane should be used (Tables 17–20).

ACKNOWLEDGMENTS

I am grateful to J. Versnel, J. M. Lesniewski, M. J. Breza, and R. Wienckoski for compounding the products and obtaining flammability and physical property data.

12. *References*

1. R. J. Schwarz, Fire Retardation of Polyethylene and Polypropylene, in *Flame retardancy of Polymeric Materials* (W. C. Kuryla and A. J. Papa, eds.), Vol. 2, pp. 83–133, Marcel Dekker, New York (1973).
2. ASTM D 2863-70, Annual Book of ASTM Standards, Part 27, p. 729, ASTM, Philadelphia, Pennsylvania (1973).
3. Standards for Tests for Flammability of Plastic Materials for Parts in Devices and Appliances, Subject 94, Underwriters Laboratory, Northbrook, Illinois, September 17, 1973.
4. J. Green and J. Versnel, *J. Fire Flammability, Fire Retardant Chem.* **1**, 185–204 (November 1974).
5. Standard for Safety, Television Receivers and Video Products, UL 1410, Underwriters Laboratories, Melville, New York (1976).
6. Recognized Component Directory, p. 460, Underwriters Laboratories, Inc., March 1977.
7. J. W. Vanderhoff and A. K. John, U.S. Pat. 3,058,929 (October 16, 1962).
8. J. DiPietro and H. Stepneczka, *SPE J.* **27**, 23 (February 1971).
9. J. J. Pitts, P. H. Scott, and D. G. Powell, *J. Cell Plastics* **6**, 35 (January/February 1970).
10. R. J. Schwartz, Fire Retardation of Polyethylene and Polypropylene, in *Flame Retardancy of Polymeric Materials* (W. C. Kuryla and A. J. Papa, eds.), Vol. 2, p. 88, Marcel Dekker, New York (1973).
11. J. A. Wiesman, *Electra* **55**, 25–38 (December 1977).
12. R. M. Eichhorn, *IEEE Trans. Electr. Insul.* **EI-12**, 2–45 (February 1976).
13. R. B. Blodgett, EP Rubber and Crosslinked Polyethylene as Insulations for 90°C Rated Medium Voltage Cables, presented at ACS Rubber Division, October 12, 1978, Boston, Massachusetts.

14. U. I. Vaidya, Flame Retarded EPDM Integral Insulation Jacket Compositions with Excellent Heat Resistance and Electrical Stability, presented at ACS Rubber Division, October 12, 1978, Boston, Massachusetts.
15. Union Carbide Co., Silane Coupling Agents in Mineral—Reinforced Elastomers, F4471B, p. 4, September 1977.
16. A. M. Gessler, Evidence for Chemical Interaction in Carbon and Polymer Associations, Rubber Chem. Tech., pp. 850–873 (June 1969).

2

Methods for Reduction of Smoke from Burning Polymers

David F. Lawson

1. Introduction

A 1973 report of the National Commission on Fire Prevention and Control cited the flammability of materials as only one of several aspects of America's fire problem, which also includes fire prevention, personal education, firefighting, medical treatment, building and systems design, and fire detection, containment, and extinguishment.[1] Materials, whether petroleum stocks, grains, lumber, or synthetic polymers, present a source of fuel for combustion. Historically, the chemical treatment of materials to improve fire resistance can be traced back at least as far as the eighteenth century, and the literature describing such treatments has since grown voluminous. However, smoke generation is one of the hazards associated with materials' flammability, and this has received attention only relatively recently.

Most of the work on smoke generation has taken place in the last decade, beginning with the development of test methods for measuring smoke density from burning materials. Ironically, no single smoke generation test is yet universally recognized for its predictive ability or correlation with actual fire situations. It has been demonstrated by several groups that both the intensity and rate of development of visible smoke can be important factors in assessing

David F. Lawson • Central Research Laboratories, The Firestone Tire & Rubber Co., Akron, Ohio.

fire survivability. The recent interest in smoke generation tendencies and the potential for related standards have stimulated industry to seek methods for reducing smoke output from those products which may become involved in fires.

The purpose of this chapter is to review the present technology for reducing visible smoke from polymeric materials. For background, an idealized solution to the problem of flame resistance and smoke inhibition in polymers as well as practical constraints to the solutions will be considered. As further background, the phenomenon of smoke and test methods for measuring smoke will be discussed. Then the present technology dealing with smoke generation and inhibition will be surveyed with emphasis on the underlying chemistry whenever known. Finally, generalizations about factors which favor smoke reduction will be made and trends for future developments will be traced. Wherever possible, the point of view of polymer technologists seeking improvements for a given polymer, will be assumed. Comparisons will be made with new polymers and those in which structural modifications have resulted in the reductions of smoke when burned. For the purposes of this chapter, the terms *flame resistant*, *flame retardant*, *fire retardant*, and *fire resistant* will be considered equivalent, as will terms such as *smoke retardant*, *smoke inhibited*, *smoke suppressed*, etc.

"Ideal" Flame-Retardant Systems

In general, the improvements sought in flame-resistant polymers encompass the entire spectrum of burning behavior, manifested as ignitability, flame spread, glow, flashover, heat contributions, visible smoke, toxic or irritating combustion gases, and oxygen depletion. Ideally, an improved material system will minimize the hazards associated with all of these while at the same time maintaining or improving other properties. At the very least, a reduction in one type of hazard should not be accompanied by the increase of another type of hazard. In assessing the relative merits of smoke inhibitor systems, the overall improvement of burning behavior must be considered. Moreover, besides improving burning behavior, an "ideal" fire retardant system will: (a) minimize adverse effects on desirable properties, such as strength, color, stability, processibility, etc., (b) have versatility across a product line and show good durability in service, and (c) cause no increase in cost, preferably lowering costs. The existence of any such "ideal" system is uncertain, but it represents a goal for development work.

2. Smoke

2.1. Phenomenon

The term *smoke* has a vague definition and several connotations. In general use, smoke is considered to be a cloud of particles, individually invisible, which is opaque due to the scattering and/or absorption of visible light.[2] *Fumes* are considered to be less opaque forms of smoke. Hilado's definition of smoke as the composite of gaseous products of burning materials in which small solid or liquid particles are also dispersed[3] relates the phenomenon to combustion and introduces the aspect of gaseous products. However, it is necessary to distinguish between "combustion gases" and "visible smoke" since the two have different sources and effects, as well as different methods of measurement and significance in fire scenarios.

Among the combustion gases, carbon monoxide, CO, has commonly been of chief concern, with several reports examining its role in detail.[4–9] Other toxic gases which can be formed in fires include hydrogen cyanide, nitrogen oxides, hydrogen chloride, and sulfur oxides. Considerable work is currently underway to study combustion gases, which are leading causes of death.[10] However, several studies of planned, full-scale fires in which both visible smoke and combustion gases are monitored have shown that intense levels of visible smoke may precede the formation of excessive quantities of toxic combustion gases.[11–15] This suggests that in certain fire situations both aspects can be of comparable concern, since the loss of visibility due to heavy smoke can hinder escape until toxic gas concentrations and temperatures become critical. Presumably, reductions in the rate or intensity of visible smoke development will help to increase escape time. In this review, only the aspect of visible smoke, which is defined as that capable of obscuring light and visibility, is considered.

Visible smoke from burning polymers is generally thought to result from incomplete combustion. Since polymer flames are diffusion flames, stoichiometric mixing for complete burning does not readily occur. The visible smoke formed can consist of liquids (e.g., mists formed by liquid pyrolyzates or the partial oxidation products of these pyrolyzates, or by water) or solids (e.g., carbon flakes or spheres, ablated ash, sublimed pyrolyzates or compounding ingredients, or solid oxides from heteroorganic substrates). Carbonaceous or aqueous components are probably most commonly encountered in polymer smokes.

Early investigations into the structure of flames and the formation of smoke include the elegant demonstrations of Faraday on the chemistry of a candle.[16] More modern investigations, aided by mass spectrometry and special kinetic methods, have shown that within a flame, acetylene molecules formed by thermal cracking of the fuel will polymerize and dehydrogenate to form carbon, or soot.[17,18] During this polymerization the intermediate molecules can either remain open in the form of polyacetylenes or they can cyclize to form polybenzenoid structures, both of which will lead to soot.[19,20] The polybenzenoid structures may take on more importance as intermediates when they can be formed directly, as from aromatic fuels.[21] More detailed reviews of the chemistry of flames and soot formation are found elsewhere.[22,23]

The processes outlined above also apply to polymers, since in the presence of a sufficiently intense heat source a polymer will pyrolyze, breaking down to low-molecular-weight species. These species diffuse from the condensed phase into the dispersed phase, where they comprise the smoke observed in the absence of flame. At high heating rates and with ignition, these low-molecular-weight species fuel the polymer flame, where solid carbon, or soot, is formed in a deficiency of oxygen. The soot-forming processes are summarized in Scheme I.[23]

Scheme I:

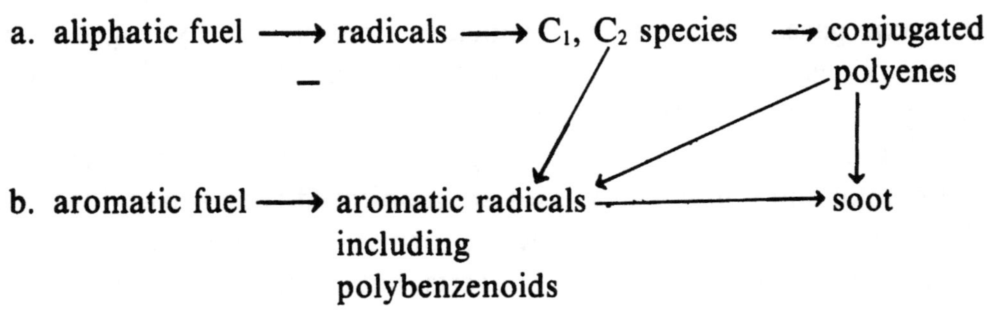

The importance of the nature of the polymer pyrolyzates in affecting soot formation is shown here. Aliphatic fuels are cracked to smaller alkyl radicals which, in a deficiency of oxidizer, grow to form conjugated polyenes or polybenzenoids which may be radical, ionic, or neutral species. Ultimately, the intermediates, which are highly reactive with other unsaturated species, condense to form soot. When acetylenic or polyenic fuels are present initially, the polyene intermediates are more readily formed, while aromatic fuels are thought to proceed directly to polybenzenoid intermediates.[24] In these cases, heavy soot formation occurs rapidly. Since oxidation to carbon oxides is competitive with soot formation, oxygen-containing fuels generally show a decreased tendency for soot formation. On the other hand, halogens may promote soot formation through dehydrohalogenation, assisting ring closure and the formation of olefins and polyenes.[24]

The relative tendency of fuels to form smoke is frequently characterized by "smoke point" values, where smoke point is defined as the "height in millimeters of the highest flame produced without smoking when the fuel is burned in a specified test lamp."[25] Smoke points of some representative compounds are listed in Table 1. These results show an obvious distinction between aliphatic, olefinic, and aromatic fuels in their propensity toward smoke formation. The relative values are consistent with the foregoing discussion, and they also show that branching favors smoke formation, presumably by increasing the generation of small radicals.

The nature of the pyrolyzates generated is thus a major determinant of smoke formation, given similar conditions of polymer combustion. Chief pyrolyzates of some common polymers are listed below:

polyethylene—ethylene, propylene, higher olefins.
poly(methyl methacrylate)—methyl methacrylate (monomer).
polystyrene—styrene, styrene oligomers, aromatics.
polyurethanes—aromatic isocyanates and amines, aldehydes.
poly(vinyl chloride)—HCl, benzene, other aromatics and low alkenes and alkanes.
neoprene—hydrogen chloride and aromatic and polyenic species.
poly(ethylene terephthalate)—acetaldehyde, unsaturated esters, carboxylated aromatics.

The relative distribution of pyrolysis products from an individual polymer is also dependent on the pyrolysis temperature and the heating rate, as well as the pyrolysis atmosphere. For example, it has been shown for a series of polyester resins that yields of styrene and toluene pass through maxima at 600–700° C,[23] and naphthalene is found among the pyrolyzates at 700° C and above. The amount of smoke generated in a nitrogen atmosphere passed through maxima with increasing temperature in several of the resins, while the smoke from others increased steadily with temperature. A study of the variation of smoke intensity with temperature for a group of eight natural and 12 synthetic polymers showed that smoke density passed through a maximum in the region of 400–600° C for all except poly(methyl methacrylate.[26]

Recent studies of smoke generation phenomena in polymers have

TABLE 1
Smoke Points of Some Organic Compounds[a]

Compound	Smoke point (mm)
n-Hexane	149
n-Octane	149
2,2,4-Trimethylpentane	86
1-Hexene	88
1-Octene	99
Cyclohexane	117
Cyclohexene	47
1-Octyne	24
Benzene	8
Toluene	6
Styrene	4
Naphthalene	4
n-Hexylmercaptan	102
Thiophenol	5
Triethylamine	122
Pyridine	29

[a]Reference 25.

concerned physical characteristics of the smoke particulates, such as shape and size distribution,[27–30] transport properties,[31] and optical effects,[32,33] as well as the combustion processes of the polymers.[34,35] Considerably more work, covering a wide range of polymer structures, compositions, and combustion conditions, is necessary before a detailed understanding of the mechanisms of smoke formation from burning polymers will be developed.

2.2. Smoke Measurement

Before the measurement of smoke can be discussed, it must be recognized that smoke generation is not an intrinsic material property, as would be the heat of combustion, melting point, or density.[36] The level of smoke actually measured in a test will depend greatly upon the burning conditions (i.e., heat flux, oxidant supply, specimen geometry, the presence or absence of flame) as well as the state of the test environment (ambient temperature, test chamber volume, ventilation, drafts, etc.). As a result, no single smoke measurement, or even a limited set of smoke measurements, is likely to provide a comprehensive definition of smoke behavior in a fire. Neither should the results from smoke tests of different design and control of variables be expected to correlate.

A fair number of tests to measure smoke generation tendencies have been developed or proposed.[23,37–49] The names and features of some of the more common tests are summarized in Table 2. As the table shows, two types of smoke measurement are generally used: light attenuation (optical) or smoke mass (weight). Optical approaches are the more common. The scale of the test methods also varies widely, from the 0.2-g sample used in the European Solvay–AFNOR smoke tube, to the 7.6-m specimen of the Steiner tunnel, ASTM E-84. In most cases, the smoke test is combined with some type of a flammability test, and the smoke test design is secondary to that of the fire test. Examples of test methods which were designed principally for smoke measurement include the NBS smoke density chamber (NFPA Method 258), Rohm and Haas XP-2 chamber, Arapahoe smoke chamber, and the Solvay–AFNOR smoke tube method.

Several smoke measurement methods have been used more extensively than others in the United States. The Rohm and Haas XP-2 chamber (ASTM D-2843) consists of a 30 × 30 × 79-cm box, in which a small specimen, usually 2.5 × 2.5 × 0.6 cm, is burned with a propane flame. The smoke is contained within the box and is measured by the absorption of a horizontal beam of light.[39] Relative material ratings are based on the maximum smoke produced (maximum light absorption) and the smoke density rating (area under the curve of light absorption vs. time, divided by total graph area over 4 min). Light absorption readings may also be converted into specific optical densities if the geometry of the system is defined,[50] but results from the XP-2

TABLE 2
Commonly Used Smoke Tests[a]

Name	Feature	Principle	Ref.
Underwriters' Laboratories tunnel (Steiner tunnel) (ASTM E-84)	7.6-m Flame-spread chamber, monitored at exhaust	Optical	37
Forest Products Laboratory tunnel (ASTM E-286)	2.4-m Flame-spread chamber, monitored at exhaust	Optical	38
Rohm and Haas Chamber (XP-2) (ASTM D-2843)	Specimen variability, flaming combustion, accumulated smoke monitored in 0.07-m^3 chamber	Optical	39
NBS chamber	76 × 76-mm specimen, radiant heating with or without flame, accumulated products monitored within 0.51-m^3 chamber	Optical	40
NBS radiant panel (ASTM E-162)	0.46-m Inclined specimen, radiant heating, piloted ignition, particulate collected at exhaust	Weight or optical	41
Cass collector	Soot from burning rod deposited on glass filter	Weight	42
Michigan Chemical densitometer	Monitors smoke rising from oxygen index test	Optical	43
Arapahoe smoke chamber	Soot collected on glass filter	Weight	47
Ohio State calorimeter	Variable heat input smoke monitored at top of exhaust stack	Optical	48
Solvay-AFNOR smoke tube	0.2-g Sample burnt in tube furnace, smoke homogenized and passed through photocell	Optical	49

[a]Reference 72.

test have most often been expressed as light absorption at a given time interval.

The XP-2 chamber evolved into the NBS smoke density chamber, which was initially used to rate materials for aircraft interiors.[40] It has become the most widely used smoke test for several reasons, including reproducibility of data and versatility, using a calibrated heat source with or without piloted ignition. The chamber consists of a 0.5-m^3 box in which a vertically mounted 7.6 × 7.6-cm specimen is burned under either flaming or nonflaming conditions. The smoke evolved is contained within the chamber and monitored with a vertical beam of light to reduce errors from stratification of smoke. Additionally, most of the combustion products are retained in the chamber, allowing for assays during the test. Smoke intensity is followed by light transmittance, and is expressed in terms of a dimensionless specific optical density, D_s, defined as

$$D_s = G \log_{10} \frac{T_o}{T} = \frac{V}{AL} \log_{10} \frac{T_o}{T} \tag{1}$$

where V is the volume of smoke chamber, A the area of burning or smoldering material, L the length of light path, T_o the initial % transmittance, and T is the measured % transmittance. For the NBS test as it is usually performed, the geometric factor, G, is 132. Based on this definition, higher D_s is indicative of greater visible smoke intensity. A maximum corrected specific optical density D_{mc}, which is compensated for smoke particles adhering to lens seals, is usually reported, as are times to given optical densities. Table 3 summarizes some of the terms used in reports of NBS smoke density test results.

The Underwriters' Laboratories (UL), or Steiner, tunnel uses a specimen 7.6 m long and 49.5 cm wide, which is placed face down at the top of the tunnel. Two gas burners are used as the ignition source, and conditions are adjusted such that a flame will spread the length of the chamber over a red oak control in 330 sec. The intensity of smoke flowing from the tunnel is measured with a photometer positioned vertically at the vent stack. The smoke generated is usually expressed relative to the total output of the red oak control, which is assigned a value of 100. Results are also sometimes reported as optical densities.

The Arapahoe smoke chamber consists of a vertical cylinder 63.5 cm high, with a filter assembly at the top. A 3.8 × 1.25 × 0.3-cm specimen is ignited at the base of the chamber with a propane burner, and smoke particles are collected by drawing a suction through the filter paper. Smoke quantities are expressed as a percentage of the initial sample weight or the weight of sample burnt.

There has been some interest in comparing and correlating various large-and small-scale laboratory smoke tests, and also in validating smoke density tests of any scale with full-scale fires. Researchers at Arapahoe found

TABLE 3
Terminology Used in NBS Smoke Density Test[a]

Term	Definition	Interpretation
D_m	Maximum specific optical density = 132 $\log_{10}(100/\%T_m)$	
D_c	Clear beam density—after test with chamber exhausted	Indicates lens fogging during test
D_{mc}	$D_m - D_c$	D_{mc}: 50, moderately dense, 100–300 dense greater than 400, very dense[b]
$t.9D_m$	Time (minutes) to 90% of maximum optical density	
tD_{16}	Time (minutes) to D = 16; obscuration time	tD_{16} less than 3, fast 4–7, moderate greater than 7, slow[c]
SON_4	Smoke obscuration number, average D_m at t = 1, 2, 3, 4 minutes	
R_m	Maximum rate of development of smoke (min^{-1})	

[a]Reference 72.
[b]Arbitrary ranges, Reference 51.
[c]Arbitrary ranges.

that the detection of smoke-suppressant effects due to a given additive was inconsistent between the Arapahoe and NBS smoke chambers and the UL tunnel.(12) Later, the same group reported that Arapahoe gravimetric smoke data and NBS smoke density chamber optical values for sheet samples burnt under flaming conditions correlated well if the burning characteristics of the material were similar. An interlaboratory comparison of 17 materials resulted in a correlation with a coefficient of 0.88 between the optical density predicted by the gravimetric method and those measured in the NBS chamber.(52) A correlation analysis has been reported which compares data from the NBS smoke density chamber and the UL tunnel with corner-room fire results for a group of wall-covering materials tested at the National Bureau of Standards.(12,53) Correlations of 81% for the UL tunnel, and 50% and 66% for the NBS chamber under flaming and smoldering conditions, respectively, were indicated. Reasons for the differences are uncertain, but may reflect differences in heat flux, geometry, or the dynamics of each system. The superior correlation of the UL tunnel in this case may, perhaps, be fortuitous. The UL tunnel can hardly be considered as a small-scale or screening procedure for smoking tendencies.

In all of the smoke tests described, the value of principal interest has been the amount of smoke formed. However, since the full-scale tests discussed earlier (in Section 2.1) indicate that fire hazards from smoke are associated with its appearance early in the fire time scale, the rate of generation of smoke must also be an important consideration. Recently proposed Federal Aviation Administration and Department of Transportation materials standards recognize this by limiting optical densities in the NBS smoke chamber to a maximum of 100 in 90 sec and 200 in 240 sec. It appears to be agreed that the fire hazard presented by visible smoke is impedance of egress from the fire site. However, it remains to be conclusively demonstrated that the optical or mass properties measured in laboratory smoke tests represent the actual fire hazard, and that other factors, such as irritation, combined with smoke opacity are not responsible for fire hazards. If other factors are involved, how are these measured? Further research appears to be warranted here.

2.3. *Factors Affecting Measured Smoke Values*

A multitude of common polymers have been studied using the test procedures described above. Because of the variety of polymer types, constructions, and test methods used, factors which influence the test results are recognized as being of two general types: (1) polymer related, including polymer structure, sample composition, and construction, and size of the test specimen; and (2) test related, including the general design of the test, ventilation, heat flux, and sample orientation.

2.3.1. *Polymer Structure*

When discernible, structural trends in smoke measurements are consistent with the contribution of fuels to smoke formation. This is apparent in results obtained with the Rohm and Haas XP-2 chamber under flaming conditions (Table 4). Here, polymers with aliphatic backbones, or those that are largely aliphatic and oxygenated, showed a tendency toward low smoke generation, while polyenic polymers and those with pendant aromatic groups generally tended toward greater smoke. There are two apparent exceptions to this in Table 4, the saturated polymers polyvinyl chloride and polyvinylidene chloride, which demonstrate that the nature and amount of combustion fuels are also important factors. In polyvinyl chloride, dehydrohalogenation gives a polyenic structure which pyrolyzes to form aromatic products, while polyvinylidene chloride forms more char and different volatile pyrolyzates. Polymers with high thermal stability or which form limited amounts of flammable pyrolyzates generally have limited visible smoke. Char formation is one conceivable mechanism for minimizing the yield of pyrolyzates.

The exact order of smoking tendencies changes somewhat when the results from the NBS smoke density chamber, vertical sample orientation, shown in Table 5, are considered.[54] However, the general groupings are

TABLE 4
Smoke Chamber Results for Various Materials
Using Rohm & Haas XP-2 Chamber[a]

Material	D_m
Acetal	0
Nylon 6	1
Polymethyl methacrylate	2
Polyethylene, low density	13
Polyethylene, high density	39
Polypropylene	41
Polyvinylidene chloride	98
Polyethylene terephthalate	390
Polycarbonate	427
Polystyrene	494
Acrylonitrile-butadiene-styrene	720
Polyvinyl chloride	720

[a]Reference 50.

TABLE 5
NBS Smoke Density Results for Various Materials[a]

Material	Thickness (mm)	D_{mc}	
		Flaming	Nonflaming
Red oak	6.3	76	395
White pine	6.3	155	325
Polyethylene	6.3	150	470
Polytetrafluoroethylene	6.3	55	0
Polyvinyl chloride	3.2	525	272
Polyvinyl chloride	6.3	660	300
Polystyrene	6.3	660	372
Polystyrene foam, F.R.	6.3	260	10
Acrylonitrile-butadiene-styrene	1.2	660	71
Acrylic	5.5	107	156
Acrylic, F.R.	6.3	480	380
Cellulose acetate butyrate	6.3	49	434
Polycarbonate	3.2	174	12
Polyurethane, flexible foam	12.7	20	156
Polyurethane, rigid foam	25.4	439	117
Natural rubber foam	19.0	660	236

[a]Reference 54.

essentially the same as those of the XP-2: under flaming conditions polyenic polymers and those with pendant aromatics are generally smokier than oxygenated or aliphatic polymers. Under nonflaming conditions these categories no longer hold, as smoking tendencies for polymers in these two arbitrary groups tend to reverse. An interpretation of smoke intensities in the NBS chamber under nonflaming conditions based primarily on structural considerations is not readily apparent. A problem with these groupings is that sheet specimens of one polymer are compared with foams of another, and thicknesses and weights vary. The direct comparison of smoke intensities such as those in Table 5, without qualifications of sample size or mass, test orientation, or burning behavior, can be misleading. Seader and coworkers have demonstrated the utility of mass-optical density, MOD, for correlating smoking tendencies with the NBS smoke density chamber.[32,55] MOD is less sensitive than optical density to differences in sample thickness and density, provided light transmittance is greater than 1%. MOD is defined as

$$\text{MOD} = \frac{D_s}{M/A} = \frac{DV}{ML} \tag{2}$$

where D_s is the specific optical density, D the optical density, M the mass loss of the test specimen, A the surface area of burning polymer, V the volume of smoke chamber, and L is the length of light path in smoke chamber. Using MOD, smoke density results are normalized to the weight lost by the materials during exposure, which is most useful for comparison of polymer structural effects.

Breden and Meisters have shown a relationship of MOD to structure (Table 6) for a limited series of carbon-backbone polymers burnt in a horizontal configuration in the NBS smoke density chamber.[56] Polystyrene, with pendant aromatic groups, exhibited the highest MOD, while the

TABLE 6

Structure and Mass-Optical Density (MOD)[a]

Material	MOD (cm^2/g)
Polystyrene $[-(Ph)CH-CH_2]_n$	14,300
Styrene $PhCH=CH_2$	9,600
Polypropylene $[-(CH_3)CH-CH_2-]_n$	5,250
PVC $[-(Cl)CH-CH_2-]_n$	3,400
Polyethylene $(-CH_2-CH_2-)_n$	2,900
Paraffin wax $(-CH_2-CH_2-)_n$	2,300
Polyoxymethylene $(-CH_2-O-)_n$	~0

[a]Reference 56.

substitution of pendant methyl, chlorine, or hydrogen resulted in progressively lower MODs. The differences between polystyrene and styrene monomer, and polyethylene and paraffin wax were ascribed to molecular weight effects. Polyoxymethylene, which lacks an all-carbon backbone, showed essentially no smoke.

Structural factors in smoke generation appear to be important, insofar as they contribute to the inherent stability of the polymer and largely determine the nature of the pyrolyzates which form the combustion fuels. The following oversimplified generalizations about structural factors may be made:

- Aromatic and polyenic polymers have greater smoking tendencies than aliphatic or oxygenated polymers.
- Polymers with aromatic units comprising the backbone have lower smoking tendencies than polymers with pendant aromatic groups.
- Halogenation to low or intermediate levels tends to increase the smoking tendency, but highly halogenated polymers have reduced smoking tendencies.
- Smoking tendencies may be related to the types of fuels formed on degradation and to the degree of thermal stability of the polymers.

For every one of these generalizations, there is certain to be an exception.

2.3.2. Specimen Construction and Dimensions

The problem of making structural comparisons from data such as those of Tables 4 and 5 readily illustrates the contributions of specimen construction and dimensions to measured smoke values. By construction we imply the physical form, density, and certain compound variations. These variables can affect heat transfer, heat and mass diffusion, and melt flow. For example, Benisek and Phillips have shown the effect of wool carpet constructions on smoke generation in the NBS chamber.[57] They found that smoke densities depended on the depth and density of the pile, as well as the identity of the backing fiber and latex. As another example, NBS smoke intensity has been shown to increase with specimen thickness in rigid PVC,[58] and a relationship between sample thickness and NBS smoke density results has been derived.[59]

Comparisons of smoking tendencies have been made on an adjusted mass (not MOD) basis. By normalizing smoke density results on the basis of equivalent "fuel" loadings (initial sample mass), Grand has developed a scale of relative "maximum hazard smoke densities"[60] Summarized in Fig. 1, these values represent the highest optical density reading for each material under either exposure condition, divided by one tenth of the sample mass. Maximum hazard smoke densities are not as useful as MOD for defining structural factors which contribute to smoke generation, but seem more useful

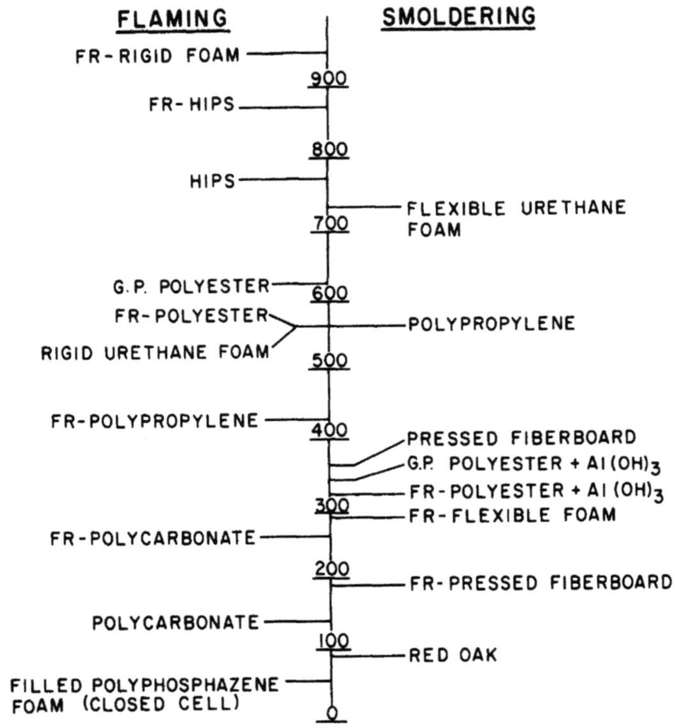

FIGURE 1. Maximum hazard smoke densities (D_m) for several polymeric materials, normalized to 10 g "fuel" (Reprinted form Ref. 60 by permission of Technomic Publishing Co., Westport, Connecticut).

for ranking potential fire hazards. Thus flexible urethane foam ranks near high-impact polystyrene on an equivalent-initial-weight basis, although the maximum smoke intensities occur under different exposure conditions. Similar rankings on the basis of equivalent volumes for materials having the same end use may perhaps be more meaningful, but some common quantity, whether mass or volume, must be the basis for comparison.

The presence of fillers or other compounding additives can also have a substantial effect on measured smoke levels. NBS smoke density levels of some thermoplastics are increased significantly when the compounds contain as little as 10% of an inert filler.[59] The increase is believed to be due to the hindrance of melt flow, forcing the exposure of more polymer to heat during the test. Other fillers, particularly at higher concentrations, may cause reductions in the amount of visible smoke, as we shall consider in the discussion on smoke inhibitors.

2.3.3. *Ventilation*

Changes in the fuel/oxidant ratio or physical removal of visible smoke components will affect smoke test results. Gaskill has used a modified version of the NBS smoke chamber to measure the effect of air changes on smoke test results.[51] His results are summarized in Table 7, showing that increased

TABLE 7
Results in Modified NBS Chamber, Showing Effect of Ventilation[a]

Material	Changes of air/hr	D_{mc}	
		F	NF
Polystyrene	0	470	460
	6	465	160
	20	465	45
Red oak	0	75	395
	6	50	285
	20	5	125

[a]Reference 51.

ventilation sweeps particulates from the chamber, decreasing optical densities. However, under flaming conditions the smoke generated from polystyrene remained high regardless of ventilation, indicating that the rate of smoke formation is faster than the ventilation rate. Sauers has also reported on forced air flow and oxygen concentration effects, using a NASA multicapability chamber.[61] In the test configuration used, maximum specific optical density was found to vary with a linear and a quadratic component for both air flow rate and oxygen concentration.

2.3.4. *Heat Flux*

The data of Tables 5 and 7 show how widely smoke densities obtained under flaming and smoldering combustion can differ. This is understandable in terms of the processes of Scheme I, wherein aerosols of pyrolysis fragments predominate in the absence of flame, while carbon particles are formed within flames. Some studies have been performed using variable radiant heat levels in the NBS smoke chamber.[62,63] Smoke intensities are known to pass through maxima with increasing heat flux in some systems.[23]

2.3.5. *Sample Orientation*

One criticism leveled against the NBS smoke density chamber has been that the vertical sample orientation allows thermoplastic materials to flow from the heated zone during a test, producing anomalously low optical densities. Breden and Meisters have examined the use of a horizontal sample configuration.[56] Their data in Table 8 show a remarkable increase in measured D_m of thermoplastics, going from a vertical to a horizontal configuration. Molten polymer can flow from the sample holder in the former case, while it is contained and burnt in a pool in the latter. Results for

TABLE 8
Comparison of Smoke Density Values, Flaming Conditions, for Vertical and Horizontal Sample Configurations[a,b]

	Average D_m	
	Horizontal	Vertical
Thermoplastic		
Polypropylene	398	57
Polyethylene	286	35
Nylon 6,6	264	48
Paraffin wax	228	83
Nonthermoplastic		
Phenolic impregnated paper	155	140
Vulcanized fiber	52	63
Balsa wood	16	8

[a]Reference 56.
[b]All materials tested in 0.15-cm thickness, NBS smoke density chamber.

nonthermoplastic materials showed little difference when tested in the horizontal or vertical position.

3. *Smoke Inhibition Technology*

3.1. *Background*

Early work on smoke-suppressant treatments was directed at additives for reducing particulate emissions from burner fuels; some of this work has previously been reviewed,[64,65] The effectiveness of these additives varies with the fuel and is dependent on concentration as well. Some generalizations about smoke-suppressant additives for fuels used in energy conversion processes can be made:

a. Additives of necessity function in gas phase flame processes.
b. The predominant class of smoke-suppressant additives for fuels are metal salts and complexes (other classes include sulfur and nitrogen oxide precursors).
c. If rankings of effectiveness were compiled, the following would be representative: most effective—barium compounds, e.g., barium carboxylates and sulfonates[66–68]; most versatile—manganese compounds, e.g., methylcyclopentadienyl manganese tricarbonyl[64]; variable effectiveness—iron compounds, e.g., dicyclopentadienyl iron (ferrocene).[69,70]

The metals appear to function as smoke suppressants through the gas phase destruction or the prevention of formation of soot particles. A general mechanism for destruction of solid carbon in the dispersed phase has been proposed (Scheme II).[71]

Scheme II:

$$MO + H_2 \rightleftharpoons MOH + H\cdot$$

$$MOH + H_2O \rightleftharpoons M(OH)_2 + H\cdot$$

$$M(OH)_2 + (X) \rightleftharpoons MO + H_2O + X$$

$$H\cdot + H_2O \rightleftharpoons \cdot OH + H_2$$

$$\cdot OH + C_{(s)} \rightleftharpoons CO + H\cdot$$

In this mechanism, oxides or hydroxides of the metals catalyze the decomposition of hydrogen or water vapor, producing hydroxyl radicals, which will oxidize solid carbon. Other mechanisms have also been postulated, in which carbon is oxidized directly by reaction with metal oxides,[71] or in which ionic intermediates or precursors of soot are neutralized by contact with metal species.[67]

3.2. *Smoke Inhibition in Polymers*

Unlike burner fuels, where smoke suppressants contribute in gas phase flame processes, the combustion of polymers involves a variety of processes occurring in several phases. Thus, polymer melting and degradation, conductance and radiation of heat, and diffusion of pyrolyzates through the degrading polymer into the dispersed phase accompany the various gas phase phenomena. As a result, polymers should lend themselves, at least in principle, to more varied smoke-suppressant treatments. Not surprisingly, perhaps, several of the smoke-suppressant additives known to be effective in burner fuels are also effective for polymers, although some are incompatible for various reasons. Other approaches to reducing smoke in burning polymers have included the use of fillers, additives, surface treatments, and structural modification of the polymers. Much of the polymer smoke inhibitor technology known prior to 1975 has been reviewed,[72] and considerably more technology has been developed since then. Little information concerning the chemical and/or physical mechanisms of smoke suppression in polymers has been published. However, certain types of processes occurring during combustion might be expected to affect the generation of visible smoke. These include the following.

Dispersed phase: oxidation and hindered nucleation of soot, flocculation of soot particles, dilution or enrichment of fuel mixtures, modification of the type or distribution of pyrolyzates, modification of flame temperatures.

Condensed phase: dilution of combustible polymer content, dissipation of heat, surface insulation or isolation of combustible substrate, promotion of char formation, alteration of pyrolysis reactions and energetics, alteration of the type or distribution of pyrolyzates, reduction of polymer mass burning rate.

For a given, effective smoke-suppressant treatment, simultaneous contributions to any number of these processes are possible.

3.2.1. Examples of Smoke-Suppressant Treatments for Polymers

3.2.1.1 Fillers. By fillers, we refer to nonpolymeric compounding ingredients used at concentrations greater than about 20% of the polymer mass, sometimes to concentrations as high as 400%. In some of the earliest known uses of polymers, fillers have been incorporated. Amazon Indians put black powder in rubber to improve light stability, and since then various compounds of rubbers and other polymers have been made with ground limestone, silica, talc, clay, barytes, kaolin, zinc oxide, carbon black, glass, sawdust, etc., to extend the materials and improve physical properties.[73] In 1921, the use of $Al(OH)_3$ in combination with $SbCl_3$ as a flame-retardant treatment for rubber was proposed.[74] This may be among the earliest, unrecognized smoke-suppressant treatments for polymers, since it has been demonstrated more recently that $Al(OH)_3$ (also written as $Al_2O_3 \cdot 3H_2O$, alumina trihydrate) not only retards flammability, but also reduces smoke generation in a number of polymers.[36,75–77] Other filler materials commonly used have also been found to reduce the formation of smoke. Although many different interactions can take place among the diverse polymer–filler combinations which are possible, we can discuss two general arbitrary categories of fillers, based on their apparent smoke-suppressant functions: "inert" and "active" smoke-suppressant fillers.

a. "Inert" fillers can serve to reduce the amount of smoke generated from a given mass or volume of a polymer compound by diluting the amount of combustible substrate and by absorbing heat to reduce the mass burning rate. Examples of fillers which might be classified as inert include silica (SiO_2), asbestos, clays, barytes, $CaCO_3$, and carbon black. Vinyl–asbestos floor tiles are a well-known example. It is possible that a filler may be inert in one polymer system, but active in another. For example, $CaCO_3$ may remain unchanged during the combustion of polypropylene, but will react with HCl formed during combustion of a chlorinated polymer to liberate water and carbon dioxide. Inert fillers usually give only marginal improvements in flame resistance, unless present at very high concentrations.

b. "Active" fillers serve the same diluent and heat absorption functions as inert fillers, but they absorb more heat per unit weight through endothermal processes such as dehydration or calcination. Furthermore, gases such as water, carbon dioxide, or ammonia, released during heating may also dilute the fuel mixtures and intervene in flame processes. Examples of active fillers include various hydrates and carbonates, such as $Al(OH)_3$, basic magnesium carbonate, $Mg(OH)_2$, talcs, hydrated magnesium oxychloride, gypsum, and dawsonite. In all of these, the water of hydration and/or carbon dioxide is released at temperatures approximating those of polymer decomposition, producing both flame-retardant and smoke-suppressant effects.

Alumina trihydrate, or gibbsite, is the chemical currently used in largest volume for flame-retardant applications. Endothermic dehydration occurs over the temperature range of 240–500° C, with the maximum rate of weight loss occurring at 250–300° C. Enthalpies of dehydration ranging from +0.28 to

$$2\ Al(OH)_3 \xrightarrow{240\text{–}500°C} Al_2O_3 + 3\ H_2O\ [(+39)\text{–}(+72)\ \text{kcal/mol}] \qquad (3)$$

+0.47 kcal/g have been reported.[78,79] Because of these properties, alumina trihydrate has shown good effect as both a flame-retardant and smoke-suppressant filler in many areas.

The data compiled in Table 9 summarize a few of the flame-retardant and smoke-suppressant applications of alumina trihydrate. Whenever possible, comparisons are made with $CaCO_3$, a supposedly inert filler. The data show the much greater flame-retardant effect of alumina trihydrate relative to calcium carbonate. Smoke generation was also reduced in most cases, although polystyrene requires a loading of greater than 40% by weight to show any smoke inhibition. Physical properties of the polymers are affected by the addition of fillers, but no more so with alumina trihydrate than with $CaCO_3$. High-impact polymers, such as A–B–S, suffer massive losses of impact strength in the presence of fillers, making filler treatments impractical in systems requiring these properties. Losses of impact and tensile properties in other polymers are said to be moderated through the addition of rubber or glass fiber to the composite.[75] Thermosetting resins and rubber carpet backing and foams, which typically use high filler loadings, are large-volume markets for alumina trihydrate.

The lack of smoke-suppressant activity from alumina trihydrate at loadings of 40% by weight and lower in polystyrene is puzzling. Sobolev and Woycheshin suggested that this was due to pyrolysis of polystyrene occurring prior to dehydration of the filler.[75] Thermogravimetric analyses of filled polystyrenes suggest the opposite: at a heating rate of 10° C/min, dehydration occurs over the range of 250–310° C, while polymer weight loss takes place in a later, distinct step at 390–450° C.[81] On the other hand, magnesium hydroxide, which undergoes dehydration in the vicinity of 350–400° C, suppresses

TABLE 9
Flame- and Smoke-Retardant Properties of Alumina Trihydrate in Selected Polymers

Polymer	$Al(OH)_3$ (wt.%)	$CaCO_3$ (wt.%)	Flammability	Maximum Smoke density, (%) (XP-2)	Smoke density rating (%) (XP-2)	D_{mc} (NBS)	Tensile strength (MPa)	Notched Izod impact (J/m)
Polypropylene[a]	0	0	0.36 mm/sec[b]	24	15	–	36.4	23.5
Polypropylene[a]	40	0	0.20 mm/sec[b]	11	6	–	22.8	19.7
Polypropylene[a]	0	40	0.36 mm/sec[b]	26	5	–	22.1	20.8
Polystyrene[a]	0	0	OI = 17.9	97	84	–	26.3	10.7
Polystyrene[a]	40	0	OI = 22.6	99	78	–	20.0	10.7
Polystyrene[a]	0	40	OI = 19.6	98	80	–	19.6	10.7
Polystyrene	0	0	OI = 19.0[c]	–	–	660(F)[d]	–	–
Polystyrene	60	0	OI = 30.0[c]	–	–	105(F)[d]	–	–
A-B-S[a]	0	0	0.64 mm/sec[b]	99	89	–	39.7	229.5
A-B-S[a]	40	0	0.32 mm/sec[b]	72	57	–	22.6	26.7
A-B-S[a]	0	40	0.52 mm/sec[b]	99	82	–	22.8	26.7
Unsaturated polyester[e]	0	0	OI = 19.0	–	–	–	–	–
Unsaturated polyester[e]	60	0	OI = 38.0	(Rel. smoke = 1)[f]		–	–	–
Unsaturated polyester[e]	0	60	OI = 20.0	(Rel. smoke = 7)[f]		–	–	–
SBR-NR latex foam[g]	53	–	OI = 23.0	–	–	185(F)	–	–
SBR-NR latex foam[g]	–	53	OI = 19.5	–	–	109(F)	–	–

[a] Reference 75.
[b] UL-94 horizontal burn rate.
[c] Reference 81.
[d] Reference 36.
[e] Reference 77.
[f] Michigan Chemical Smoke Densitometer, Reference 43.
[g] Reference 80.

smoke from polystyrene when used at a loading of 40 wt.%. Thermogravimetric analyses showed dehydration and polymer weight loss occurring simultaneously in this system, which indicates that the relative timing of dehydration and depolymerization is important to the smoke inhibition efficiency of hydrated fillers. More recently, Sobolev and Woycheshin have reported that ammonium dawsonite, which dehydrates over the range 170–230°C, about 50°C lower than alumina trihydrate, also reduces smoke generation from polystyrene.[82] Factors in addition to the timing of pyrolysis and hydrate release may be involved.

Magnesium hydroxide ($MgOH_2$) has also been found to be an effective smoke inhibitor in several elastomers[80,83,84] and polyester coatings.[85] In certain styrene–butadiene rubber foams, the smoke-suppressant effect of $Mg(OH)_2$ is superior to that of $Al(OH)_3$.[80] Table 10 shows a comparison of oxygen indices, (OI), visible smoke, smoke mass and char residue for rubber foams containing 53% by weight of $CaCO_3$, $Al(OH)_3$ or $Mg(OH)_2$. These data indicate the effectiveness of both $Al(OH)_3$ and $Mg(OH)_2$ in reducing smoke and increasing OI. The hydrates appear to function in the condensed phase through their endothermicity, favoring cross-linking (char-forming) reactions over pyrolysis in butadiene–styrene rubber. $Mg(OH)_2$, which dehydrates at $\geqslant$350°C, is apparently more compatible with the temperature region of decomposition of rubber than is $Al(OH)_3$.

Fillers, both inert and active, have been employed in a wide range of polymers. Where adaptable, active fillers, notably hydrates, represent a desirable approach: in many cases they show combined flame-retardant and smoke-suppressant activity, they promote charring in some polymers, of themselves they contribute no additional toxic by-products (although interactions with the decomposing substrate are possible), and they may contribute to synergisms when used in the presence of other additives, as we will discuss later. The hydrated fillers are also generally economical: $Al(OH)_3$ in various grades ranges from $0.09 to $0.30/lb, with clays less expensive, and magne-

TABLE 10
Comparison of Fillers in SBR Foam[a]

Filler	OI	Smoke mass (%)[b]	Area of smoke curve (cm²/g)[c]	Wt% Char in Partially burnt residue
None	18.5	20.1	–	–
$CaCO_3$	18.5	7.7	42.6	1
$Al(OH)_3$	24.0	4.7	13.6	5
$Mg(OH)_2$	24.0	2.4	5.2	11

[a]Reference 80.
[b]Smoke collected on glass fiber filter; see Ref. 80.
[c]Michigan Chemical Smoke Densitometer; see Ref. 43.

sium carbonate, magnesium hydroxide, and some others more expensive. As indicated earlier, there are some problems associated with the use of fillers: (1) High loadings are generally required, and not all polymer formulations will accept high levels of fillers; (2) fillers will pigment the composition; (3) physical properties, such as impact strength, tensile and elongation, and electrical properties, such as resistivity, are usually adversely affected; (4) some, such as asbestos and other fibrous minerals, are toxic or suspected to be toxic.

3.2.1.2. Additives. By additives, we generally refer to nonpolymeric compounding ingredients, typically used at levels below about 20% of the polymer mass. Exceptions would be plasticizers, processing oils, and similar additives which may be used at higher levels. A number of different types of additives have been shown or claimed to have smoke-suppressant effects. By far the largest class of examples of smoke-suppressant additives includes the various metals and metal compounds claimed for use in polyvinyl chloride (PVC) and other halogen-containing polymers. Nonmetallic examples include dicarboxylic acids, sulfur, and various plasticizers and surfactants. Several of each type are discussed in more detail in the following sections.

a. *Metal-Containing Additives.* As noted, most of the known smoke-suppressant additives for polymers, like those for burner fuels, are metal compounds. Presumably, in some cases they can fulfill the same function of soot destruction, but the polymer additives can serve other roles as well, such as promoting char formation and modifying pyrolysis processes in the condensed phase. Some of the more effective, or more commonly used, additives include the following.

1. *Antimony compounds.* Antimony oxides, e.g., Sb_2O_3, Sb_2O_4, Sb_2O_5, are the most commonly used additives for halogen-containing polymers. They are highly effective synergists for the flame-retarding properties of the halogens,[86] but are not typically considered as smoke-suppressant additives because their effect on smoke is variable. For example, Sb_2O_3 can have either no effect or reduce smoke in rigid PVC, while it can substantially increase smoke in flexible PVC or in rubber containing chlorinated paraffins. Much of antimony's activity during combustion is centered in the gas phase and is dependent upon its conversion to chlorides for volatility.[65,86–89] Several commercial, modified antimony oxides have been introduced which claim to reduce smoke generation in PVC. These are M&T's S-711® and NL's Oncor 75 RA® and Oncor 75 RAZ.® The latter two are antimony oxides extended with silica, while the last also contains zinc. Schwarcz has compared the effect of these three in a flexible PVC compound which already contains some antimony oxide.[90] The results, summarized in Table 11, showed improved smoke generation under nonflaming conditions only in the case of the zinc-modified product.

TABLE 11
Comparisons of Antimony-Based Additives in PVC[a,b]

Additive	Conc. (phr)	OI	NBS smoke density–nonflaming D_{mc}	tD_{16}	$t.9D_m$ (min)
None	0	27.1	438	1.4	10.1
Thermoguard S711	3	28.0	443	1.3	8.6
Thermoguard S711	6	28.9	445	1.4	9.7
Oncor 75RA	3	27.4	433	1.4	11.3
Oncor 75RA	6	27.6	454	1.3	9.1
Oncor 75RAZ	3	27.1	392	1.0	6.9
Oncor 75RAZ	6	27.1	325	0.8	6.8

[a]Reference 90.
[b]Recipe (phr): PVC (100), dioctyl phthalate (67), ESTYNOX 203 (3),[c] Mark M (2),[c] antimony oxide (6), additives (x), $CaCO_3$ (25-x).
[c]ESTYNOX is a trademark of Baker Castor Oil Co. and Mark is a trademark of Argus Chemical Co.

2. *Iron compounds*. Ferrocene (dicyclopentadienyl iron) was one of the first additives for which both flame retardancy and smoke inhibition were claimed.[47,91,92] When used in PVC at concentrations as low as 0.4% by weight, ferrocene significantly reduced D_{mc}, increased oxygen index, and enhanced charring, as shown in Table 12. Because of the increased charring in PVC, involvement in the condensed phase was suspected, although evidence for both condensed-phase and gas phase activity has been shown.[93] The observation of a ferricenium salt at the onset of degradation of PVC samples containing ferrocene suggested the involvement of ferricenium or its decomposition products in an autocatalytic dehydrochlorination–cross-linking process.[93] More recently, $FeCl_2$ has been identified as a product of ferrocene degradation in the presence of air and HCl at low temperatures, while ferric oxide and hydroperoxide, along with ferric chloride, are found at higher

TABLE 12
Smoke Density and Oxygen Index of Rigid PVC Samples with Ferrocene[a]

Polymer	Additive (wt.%)	NBS smoke density–flaming D_{mc}	$t.9D_m$ (min)	tD_{16} (min)	OI	Char (wt.%)
PVC, $\overline{M}_w = 4.1 \times 10^4$	0	410	2.8	0.5	42.9	9.8
PVC, $\overline{M}_w = 4.1 \times 10^4$	0.4	310	2.5	0.5	49.5	15.6
PVC, $\overline{M}_w = 1.1 \times 10^5$	0	417	2.8	0.4	44.9	12.9
PVC, $\overline{M}_w = 1.1 \times 10^5$	0.4	300	2.4	0.6	52.5	15.7

[a]Reference 93.

temperatures.[94,95] Ferrous, ferric, and ferricenium chlorides were all shown to suppress the formation of benzene and smoke from PVC samples.[95] Charred residues of PVC treated with ferrocene contain substantially less than the theoretical amount of iron, with at least 75% of the initial iron lost through volatilization.[93,95] The mechanism outlined in Scheme III has been proposed to explain these and other observations in the PVC-ferrocene system.[95] It may also serve as a prototype for other metal systems.

Scheme III:

$$\text{PVC–Cp}_2\text{Fe} \xrightarrow[500°\text{C}]{\text{air}} \text{char (75\% Fe volatilized)} \quad (4)$$

$$\text{Cp}_2\text{Fe} + \text{HCl} \xrightarrow[<300°\text{C}]{\text{air}} \underset{\text{I}}{\text{FeCl}_2} + \underset{\text{II}}{\text{Cp}_2\text{Fe}^+\text{Cl}^-} + \underset{\text{III}}{\text{FeCl}_3}\text{ (trace)} \quad (5)$$

$$\text{Cp}_2\text{Fe} + \text{HCl} \xrightarrow[500°\text{C}]{\text{air}} \underset{\text{IV}}{\alpha\text{-Fe}_2\text{O}_3} + \underset{\text{V}}{\gamma\text{-FeOOH}} + \text{III} + \text{Cp}_2\text{Fe (trace)} \quad (6)$$

Species:	*Suspected function*:
I	Cross-linking catalyst, benzene suppressant
II, III	Dehydrochlorination catalysts
IV, V	Oxidation catalysts

According to this mechanism, the various species formed in the PVC–ferrocene (Cp_2Fe) system can fulfill different functions during combustion. Thus, $FeCl_2$ (I) is thought to cross-link the polymer and suppress benzene formation during heating by decreasing the length and mobility of polyene sequences, possibly through transfer of the triplet state energy of the activated π systems. $FeCl_3$, **III**, and $Cp_2Fe^+\ Cl^-$, **II**, are Lewis acids, capable of catalyzing dehydrochlorination, while α-Fe_2O_3, **IV**, and γ-FeOOH, **V**, are strong oxidation catalysts for pyrolyzates, char, and smoke. Ferrocene itself serves primarily as a stable source of iron. Presumably, other forms of iron should show similar effects.

Offered commerically as FE-55® by Arapahoe Chemicals, ferrocene is limited in some applications by its high vapor pressure and color, and it is not recommended for use in flexible PVC. However, two other products, DFR-100® and DFR-121®, presumably based on ferrocene chemistry, are said to be applicable in flexible vinyls, adhesives, and rigid polyurethanes.[96,97]

Other iron compounds claimed as smoke inhibitors have included iron powder in PVC,[98] ferric oxide or hydroxide in vinyl chloride copolymer foams,[99] and ferrous oxalate in chlorinated polyesters.[100] Some uses of iron in combination with other metals are discussed below.

3. *Molybdenum compounds*. Molybdenum compounds have been used in flame-retardant treatments for cellulosic textiles for some time.[101] More

recently, smoke-suppressant properties have been realized with molybdenum oxide when used in PVC, usually in combination with some other metal.[102,103] These results have since been extended into other molybdenum salts in PVC and chlorinated polyesters. Table 13 summarizes oxygen index and smoke generation data for an experimental semirigid PVC cable formulation containing MoO_3, Sb_2O_3, or ammonium octamolybdate.[104] A significant smoke reduction is indicated with MoO_3 and ammonium octamolybdate, although the flame-retardant effect of MoO_3 is slightly less than that of Sb_2O_3. By using MoO_3 as a partial replacement for Sb_2O_3, smoke and flammability properties can be optimized. Table 14 shows similar results for a chlorinated polyester thermoset resin system.[104] Here, flammability and smoke properties were optimum with a combination of MoO_3 and $Al(OH)_3$. The various ammonium molybdate salts also show smoke-suppressant effects. Although molybdenum might be expected to behave like antimony in its combustion chemistry, less than 10% of the molybdenum initially present is volatilized on combustion of PVC containing MoO_3, while greater than 90% of starting Sb_2O_3 is lost under the same conditions.[105a] This suggests either that Mo functions as a smoke suppressant and flame retardant in the condensed phase, or that the volatile fraction is highly efficient in its functions. Recent experimental results indicate that MoO_3 funcations as a Lewis acid, and reduces the amount of benzene formed in precombustion processes,[105b] while other results suggest involvement in reductive coupling.[105c] Various molybdenum compounds have been reported to give smoke suppression in polymers other than PVC, but halogen seems to be a necessary coingredient. Some of the effective molybdate salts are listed in a later section.

4. *Zinc compounds.* More recently than either iron or molybdenum, zinc compounds have been shown to impart smoke-suppressant characteristics, largely in chlorinated polymers. Like many of the fillers discussed above, zinc compounds have been commonly and extensively found as ingredients in polymer formulations, in uses from vulcanization accelerators for rubber to thermal stabilizers for vinyl polymers. Table 15 lists smoke density results for flexible PVC formulations containing Sb_2O_3 and various different sources of zinc, showing that virtually any compound which will generate ZnO is a smoke suppressant for PVC.[106] Additive levels are recommended such that no greater than about 1 phr (parts per hundred resin) of ZnO is used or generated. It was also noted that many of the zinc compounds may have unfavorable effects on polymer stability.[106] Smoke-suppressant effects appear to be optimum at 1 phr or lower. Several proprietary, commercial formulations, apparently based on the smoke-suppressant properties of zinc, have been offered by NL Industries: Ongard I[107] and Ongard II.[108,109] Ongard II is said to replace Ongard I, having less moisture sensitivity. Synergistic flame-retardant and smoke-suppressant effects are found in PVC formulations when used in combination with antimony oxide.[109] Some

TABLE 13
Oxygen Index and Michigan Chemical Smoke Density Results for Semirigid Cable Jacket[a,b]

Flame retardant	Level (phr)	OI	Michigan Chemical smoke density: Relative area under smoke curve	Michigan Chemical smoke density: Percent Reduction
None	0	29.5	12.1	–
Sb_2O_3	1	30.5	11.7	9
Sb_2O_3	2	31.0	13.0	7[c]
Sb_2O_3	3	33.0	10.2	16
MoO_3	1	29.5	2.8	77
MoO_3	2	31.0	2.9	76
MoO_3	3	31.5	1.8	85
$Sb_2O_3 + MoO_3$	1 + 1	32.5	3.8	69
Ammonium octamolybdate	2	31.0	4.3	64

[a]Reference 104.
[b]Recipe (phr): PVC (100), dioctyl azelate (25), epoxy soya (2), $Al(OH)_3$ (5), tribasic lead sulfate (7), flame retardants.
[c]Increase.

TABLE 14
Oxygen Index and Michigan Chemical Smoke Density Properties of Thermoset Resin[a,b]

Flame retardant	Level (wt.% of resin)	OI	Michigan Chemical smoke density: Relative area under smoke curve	Michigan Chemical smoke density: Percent Reduction
None	0	33.0	8.4	–
Sb_2O_3	1	34.5	14.0	67[c]
Sb_2O_3	3	42.0	16.2	93[c]
MoO_3	5	36.0	5.4	36
MoO_3	7	39.0	4.3	49
$Sb_2O_3 + MoO_3$	1 + 1	39.5	3.9	54
$Sb_2O_3 + MoO_3$	1 + 2	40.0	7.1	16
Ammonium octamolybdate	5	36.5	2.7	68
Ammonium dimolybdate	5	34.0	2.6	69
$Al(OH)_3$	15	35.0	4.3	49
$Al(OH)_3 + MoO_3$	8 + 2	40.0	1.2	86
$Al(OH)_3$ + ammonium dimolybdate	8 + 2	38.5	1.0	88

[a]Reference 104.
[b]Recipe (wt.%): Atlac 711-05A resin (100), fiberglass mat (25), additive (as indicated); 1/8-in.-thick samples.
[c]Increase.

TABLE 15
Zinc Compounds as Smoke Supressants in Flexible PVC[a,b]

Additive	Concentration		NBS smoke density–flaming
	Additive (phr)	As ZnO (phr)	D_m
None	–	–	461
ZnO	0.5	0.5	336
ZnO	0.75	0.75	286
Zn dust	0.4	0.5	407
Zn phosphate tribasic	1.12	0.75	361
Zn carbonate	0.69	0.5	411
Zn sulfate, dibasic	1.01	0.75	297
Normal Zn Trimellitate	1.51	0.5	375
Acid Zn trimellitate	2.02	0.5	445
Zn borate	1.31	0.5	349
Zn peroxide	0.85	0.75	325
Zn benzoate	1.92	0.5	327
Zn caprate	2.55	0.5	378
Zn fumarate	1.13	0.5	432
Zn fumarate	1.66	0.75	376

[a]Reference 106.
[b]Recipe (phr): PVC (100), dioctyl phthalate (50), tin stabilizer (1.5), Sb_2O_3 (3), zinc additive, as ZnO (0.5–0.75).

examples of apparent synergistic effects on flammability and smoke generation have also been recognized in the use of zinc compounds and hydrated fillers in chlorine-containing polymers.[110–112] Table 16 shows data for combinations of ZnO or $2ZnO \cdot 3B_2O_3 \cdot 3.5\ H_2O$ (Firebrake ZB®, U.S. Borax Chemicals) with $Al(OH)_3$ in PVC and chlorinated polyethylene. Surprising and substantial increases in oxygen index values accompanied smoke reductions, particularly in the case of rigid PVC.

The chemical or physical mechanisms through which zinc compounds reduce smoke from chlorinated polymers have not yet been described. Processes analogous to those of ferrocene in PVC may be envisioned. Zinc chloride has been reported to reduce the formation of benzene during the pyrolysis of PVC, although not to the extent achieved with iron compounds, and smoke generation is also reduced.[95]

5. *Compounds of other metals.* The foregoing examples cover only the most common classes of metal-containing smoke-suppressant polymer additives. There are reports of many other metal-containing compounds showing smoke-suppressant properties in polymers, covering most of the useful metals, notably Al, Bi, Ca, Co, Cr, Cu, Fe, Mg, Mn, Mo, Ni, Pb, Sb, Sn, Ti,

TABLE 16
Combinations of Zinc Compounds and Alumina Trihydrate in Chlorine-Containing Polymers[a]

Polymer	Additive (phr)	NBS smoke density–flaming	
		OI	D_{mc}
PVC, rigid[b]	None	45.1	357
	ZnO, 5	50.2	317
	Al $(OH)_3$, 30	49.0	284
	ZnO, 5 plus Al $(OH)_3$, 30	70.9	187
Chlorinated polyethylene[c]	None	27.0	707
	Firebrake ZB,[d] 5	28.0	685
	Al $(OH)_3$, 30	27.5	470
	Firebrake ZB, 5 plus Al $(OH)_3$, 30	30.5	347

[a]References 110 and 111.
[b]Reference 110, recipe (phr): PVC (100), calcium stearate (2), dibutyltin bis(laurylmercaptide) (3.75), F.R. and S.S. additives as indicated.
[c]Reference 111, recipe (phr): Chlorinated polyethylene, 36% Cl (100), calcium stearate (2), F.R. and S.S. additives as indicated.
[d]Firebrake ZB is a trademark of U.S. Borax Chemical Co.

W, Zn, and Zr. Variable real or apparent effects are observed when compounds of these metals are used individually in a broad range of polymer systems. The numerous examples of these will not be discussed in detail in this section, but are listed later in the summary of smoke-suppressant treatments categorized by polymer type.

6. *Mixed-metal additives.* In most of the examples discussed above, the

TABLE 17
Combinations of Zinc and Molybdenum for Smoke Supression in PVC[a,b]

Additive	Level (phr)	NBS smoke density, normalized for sample weight	
		D_{mc}	% Reduction
None	–	350	–
ZnO	0.5	225	35.7
MoO_3	0.6	263	24.8
ZnO + MoO_3	0.5 + 0.6	205	41.4
$ZnMoO_4$	1.5	181	48.3

[a]Reference 117.
[b]Recipe (phr): PVC (100), di-2-ethylhexyl phthalate (40) Mark 275® stabilizer (Argus Chemical Co.) (2), additives as indicated.

metal compounds are added alone or in combination with antimony oxide or alumina trihydrate. However, very strong smoke-suppressant effects have been achieved in a number of cases by combining two or more metals in an additive system. The first examples of this were in patents claiming the use of iron powder in combination with cuprous oxide or molybdic oxide.[102,103] Subsequently, patents were issued claiming potassium zinc ferrocyanide, $K_2Zn_3[Fe(CN)_6]_2$, cuprocyanide, $KZnCu(CN)_4$, and similar salts in PVC.[113,114] With these salts, which combine the activity of zinc and either copper or iron, reductions in D_{mc} on the order of 50%, and marked increases in OI were reported using levels of 0.5–5 phr. Unfortunately, the combustion toxicity of PVC treated with these salts is also said to be increased, possibly because of the cyanide. Other salts combining zinc with another metal have been reported to give smoke suppression; examples of these include zinc titanate, zirconate,[115] and tungstate[116] in PVC.

Zinc and molybdenum form a particularly effective and interesting two-metal smoke-suppressant system for PVC. Table 17 shows data for combinations of ZnO and MoO_3 as well as $ZnMoO_4$ in PVC.[117] Both Sherwin-Williams Chemicals with the KemGard™ 911 series and Climax Molybdenum Co. with the Moly FR™ series, offer proprietary additives based on zinc molybdate and related systems. Table 18 contains data comparing the effects of these proprietary additives with those of molybdic oxide and antimony in flexible PVC.[118] The mixed systems should show economic advantages over the use of MoO_3, since the price of the zinc salts is less than

TABLE 18
Flammability and Smoke Formation of Semi-rigid PVC Containing Additives[a,b]

Additive (phr)	OI[c]	Arapahoe smoke data: Percent Smoke	Arapahoe smoke data: Percent reduction[d]	NBS smoke data-flaming[e]: D_{mc}	NBS smoke data-flaming[e]: Percent reduction[d]
None	28.0	9.4	–	374	–
Sb_2O_3; 3 phr	33.0	10.8	–	721	–
MoO_3; 3 phr	30.5	7.0	35	322	55
$Sb_2O_3 + MoO_3$; 1.5 phr + 1.5 phr	32.0	7.5	30	486	33
Moly FR-10®; 3 phr	31.0	5.1	52	–	–
Moly FR-21®; 3 phr	28.0	5.5	49	–	–

[a]Reference 118.
[b]Recipe (phr): PVC (100), linear alkylphthalate (40), $Al(OH)_3$ (30), lead stabilizer (5), additives as indicated.
[c]OI determined on 1.78mm specimen
[d]Compared to sample with 3 phr Sb_2O_3.
[e]Smoke density determined for 1.02mm slab.

half that of MoO_3. Sherwin-Williams reports that free zinc oxide is minimized in KemGard 911C, improving the processibility and stability in PVC, and claims favorable results in comparisons with ferrocene, barium metaborate, and Ongard II.[119] An extensive series of two-metal additive combinations for PVC which significantly reduce smoke generation has been disclosed in recent patents. Many of these involve the combination of a nickel compound with compounds of metals ranging alphabetically from aluminum to zinc. Table 19 shows examples of a few of the numerous combinations which are conceivable; a full listing is given in a later section. In the examples shown here, smoke reductions on the order of 60–100% are obtained from the combinations $NiCO_3$–ZnO,[120] NiO–CuSCN,[121] and NiO–Fe_2O_3.[122] Few other treatments have shown effects of this magnitude. Synergism has been shown in many, but not all, of these systems. Figure 2 illustrates the synergistic behavior of NiO–CuSCN and NiO–Fe_2O_3. The solid, straight lines of Fig. 2 represent D_m/g predicted if the systems were to give a linear, additive response, while the dashed, curved lines represent the type of behavior actually observed. The components may suppress smoke individually, but they are synergistic when used in combination. New synergistic systems in addition to these have been disclosed and more can be expected. There are no reports yet on the chemistry of these mixed-metal systems. The mechanism of Scheme III may possibly be

TABLE 19
Smoke Formation by PVC Formulations Containing Various Two-Metal Systems[a,b]

Additive A (phr)	Additive B (phr)	Smoke area per gram of sample[c]	Percent reduction	Ref.
None	None	67.4	–	
$NiCO_3$ (5)	–	45.9	31.9	120
–	ZnO (5)	79.8	(18.4)[d]	120
$NiCO_3$ (4)	ZnO (1)	28.0	58.2	120
NiO (10)	–	14.9	77.9	121
–	CuSCN (10)	16.7	75.2	121
NiO (5)	CuSCN (5)	5.4	92.0	121
NiO (8)	CuSCN (2)	0	100.0	121
NiO (2)	–	60.7	9.9	122
–	Fe_2O_3 (2)	24.9	63.1	122
NiO (1)	Fe_2O_3 (1)	17.6	73.9	122
NiO (5)	–	32.0	52.5	122
–	Fe_2O_3 (5)	32.8	51.3	122
NiO (4)	Fe_2O_3 (1)	8.8	86.9	122

[a] References 120–122.
[b] Recipe (phr): PVC (100), Additives A and B as indicated.
[c] Goodrich "Smoke-Char Test," measuring attenuation of light by smoke rising from sample burned for 60 sec in propane flame.
[d] Increase.

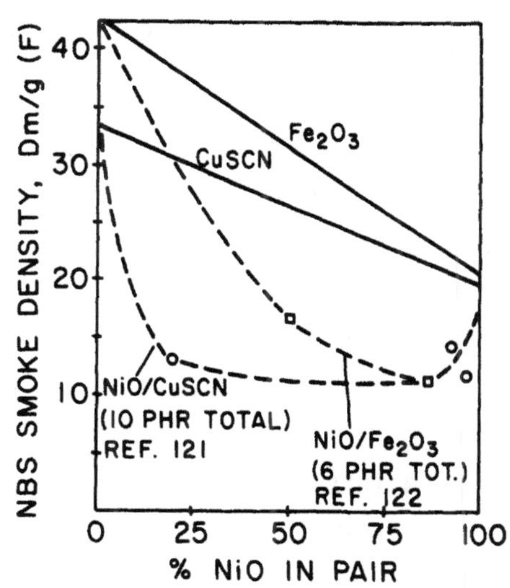

FIGURE 2. Projected smoke response curves for metal systems, illustrating synergism with NiO in PVC.[121,122]

relevant, where each of the additives in the mixed system may be efficient in a different function. An alternative speculation is that one component may serve to regenerate the second in some manner, analogous to Ba–Cd systems used in the thermal stabilization of PVC.

Recently, Kroenke has presented a survey of the effects of metal additives as smoke suppressants in PVC.[123] Using the metals individually, the members indicated in the following groups were found most effective: alkali metals, none; alkaline earth metals, Mg, $Mg(OH)_2$; lanthanides, CeO_2; posttransition metals, CdO, Zn; Group III-A, α-Al_2O_3; Group IV-A, SnO, SnO_2, $SnSO_4$, Hisil 233; Group V-A, BiOCl, $Bi_2O_2CO_3$, Bi_2O_3; Group IV-B, TiO_2, $Zr(acac)_2$; Group V-B, $VO(acac)_2$, V_2O_5; Group VI-B, $Cr(acac)_3$, Mo, W; Group VII-B, $MnCl_3$, Mn_2O_3; Group VIII, Fe, $Co(acac)_2$, $CoCO_3$, Ni; I-B, Cu. Activity was said to increase with higher atomic number within a group or in going to groups of higher order. All of the effective metal additives were found to increase char, and the additive concentration was a particularly important factor with MoO_3 and ZnO. Contributions of stabilizers were noted: the use of Sn systems was desirable when SnO_2 or ZnO were added, but interfered in the presence of $FeSO_4$ and MoO_3. Factors recognized to be important included the metal cation and its environment (cations with multiple oxidation states are preferred, although the initial oxidation state is relatively unimportant), particle size and dispersion, concentration, and synergism.

It should be noted that some of the metals showing smoke-suppressant activity have already been used for some time in polymer formulations, although for different reasons. Stabilizers for PVC based on Ba, Cd and Zn are well known, and tin-based compounds have been used as both stabilizers for PVC and activators for polyurethanes. Interferences with compounding ingredients may also arise to neutralize the effects of some smoke-suppressant additives. For instance, ferrocene is reportedly ineffective in flexible PVC with phosphate plasticizers,[47] and some PVC stabilizers reduce the effectiveness of metal-containing smoke-suppressant additives (e.g., tin stabilizers with iron additives).[123,124]

b. *Carboxylic Acids, Aldehydes, Alcohols.* One of the earlier smoke-suppressant additive treatments reported was the incorporation in a polyurethane of a solid dicarboxylic acid, such as fumaric,[125,126] isophthalic,[127] or chlorendic acid.[128] Table 20 contains an example of the use of fumaric acid at a fairly high loading in a polyether–MDI polyurethane.[125] Both flammability and smoke generation were apparently improved in the presence of the dicarboxylic acid. Flame retardancy can also optionally be increased by use of a phosphorus additive in combination with the dicarboxylic acid. The added dicarboxylic acids do not react with the urethane ingredients, but remain as discrete particulate clusters in the foam. The mechanism of smoke suppression has not been clarified, although it has been suggested that the acids may promote char formation or cross-linking among the decomposition products, or they may volatilize and alter flame processes.[126] Reactions between RCOOH and –NCO liberated by pyrolysis are also a likely possibility. Other polymer systems in which polycarboxylic acids have shown smoke-suppressant activity include polyester thermosets, flexible PVC, and polyphosphazenes. Other dicarboxylic acid derivatives, such as anhydrides and acid halides, have also been claimed to reduce smoke from rigid urethane foams.[129]

Various aromatic alcohols and aldehydes have been found to reduce smoke generation from isocyanurate foams when used at a concentration of about 10% (based on the polyisocyanate).[130] Furfural, benzaldehyde, cinnamaldehyde, furfuryl alcohol, and benzyl alcohol are examples of these

TABLE 20
Rigid Polyurethane Foam with Solid Dicarboxylic Acid Smoke Suppressant[a]

	Foam A	Foam B
p,p'-Diphenylmethane diisocyanate (parts)	90.0	90.0
Fumaric acid (parts)	–	32.0
Polyether polyol (parts)	63.6	63.6
Silicone surfactant (parts)	1.0	1.0
Dibutyl tin diacetate (parts)	0.3	0.4
$CFCl_3$ (parts)	29.1	35.0
Foam quality	Good	Good
Density (kg/m^3)	31.08	30.92
Flammability–ASTM D1692	Burns completely	Extinguished
NBS smoke density, D_m	121	86

[a]Reference 125.

additives. The smoke-retardant mechanism through which these additives function may also be relevant to the function of the dicarboxylic acids. The sequence of Scheme IV was proposed to explain the smoke suppressant activity of the alcohols and aldehydes.[130]

Scheme IV:

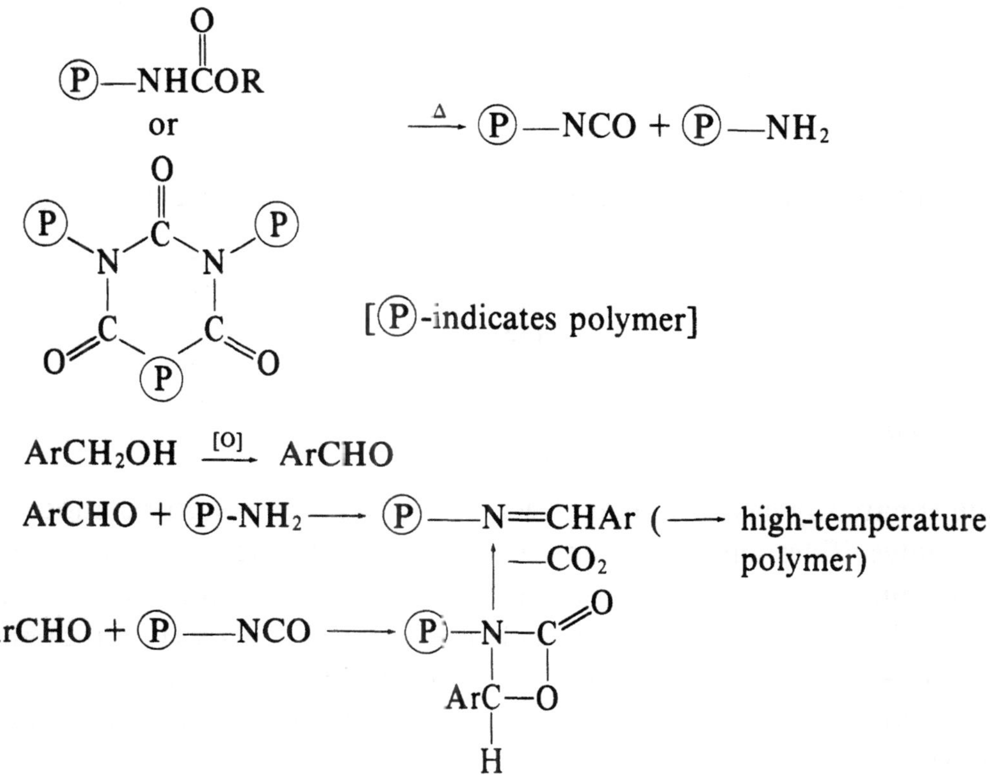

(7) (8) (9) (10)

In this scheme the aldehyde traps polymeric isocyanates or amines, which are formed upon pyrolysis of the urethane and isocyanurate structures, ultimately leading to the formation of thermally stable resin and char in the degrading polymer, and diversion of otherwise smoky aromatic fuels into solid phase by-products.

c. *Fluoroborates*. Polyisocyanurate-type foams have also been treated with alkali metal and ammonium fluoroborates to reduce smoke generation.[131] Table 21 lists some data indicative of the levels and improvements achieved by this approach. Although smoke generation was reduced, flame spread properties were not affected. The function of fluoroborate additives has not been defined, although condensed-phase catalytic involvement is a likely possibility.

d. *Sulfur*. In the discussion of smoke-suppressant treatments for burner fuels it was indicated that sulfur oxide and nitrogen oxide precursors were effective. The use of 3–20% of elemental sulfur as a smoke-suppressant additive for polyurethane foams has been claimed.[132] Approximately a 60% reduction in smoke intensity in a small tunnel test was reported when 10 phr of sulfur was added to a test formulation. Other sources of sulfur or sulfur

TABLE 21
Smoke Inhibition in Polyisocyanurate Foams Containing Fluoroborate Salts[a]

A. XP-2 smoke density chamber

Additive (wt.%)	Absorbance at 0.5 min
1. None	60%
KBF_4 (5)	53%
KBF_4 (10)	35%

B. Monsanto 2-ft tunnel results

Additive (wt.%)	Smoke index	Flame-spread index
1. None	150	25
KBF_4 (10)	65	25
2. None	165	29
NH_4BF_4 (10)	95	28

[a]Reference 131.

dioxide apparently have a similar effect, as various organosulfur compounds such as polysulfides and thiocarbonates were also reported to reduce smoke without improving flame resistance in polyurethanes and polyisocyanurates.[133] No rationale was proposed to explain this activity, although it could conceivably involve gas phase soot destruction by SO_2, condensed-phase reactions of sulfur ions or radicals, or effects on melt rheological properties.

e. *Miscellaneous Additives*. Various other additives have been claimed to reduce smoke generation. Isolated examples include the use of ammonium sulfamate in phenolic foams,[134] boric acid, borax, and silica in thermoset polyesters,[135] alkyl borate esters in phosphate-plasticized PVC,[136] and borates, phosphate salts, and esters in polyurethane foams.[137]

The substitution or modification of commonly used compounding ingredients might have potential as a means of improving smoke properties. For example, much of the smoke generated during the burning of flexible PVC is derived from the plasticizer. The substitution of aliphatic or polyether plasticizers for the more commonly used phthalates seems to be a likely solution. However, it has been shown that the presence of PVC may cause a "leveling effect" on the smoking tendencies of plasticizers, with plasticizers which burn cleanly as pure liquids smoking considerably in PVC compounds.[138] Nevertheless, dioctyl azelate[139] and dioctyl oxalate[140] have been reported to improve smoke generation when used in place of phthalates in flexible PVC. Smoke generation by flexible PVC compounds containing alkyl, aryl phosphate plasticizers is reportedly minimized when the compounds are filled with basic magnesium carbonate.[141,142]

Surfactants based on organosiloxanes and organosilicates (for example, a mixture of tetraethyl silicate and a siloxane surfactant) have been used in the preparation of rigid polyisocyanurate foams for which low smoke evolution, flame resistance and reduced friability are claimed.[143] The role of silicon in these systems was not stated. Other rigid polyurethane foams showing increased char and reduced levels of visible smoke were prepared by incorporating up to 70% by weight of powdered cellulose.[144]

It is apparent from the above discussions that although metal compounds constitute the majority of examples of smoke-inhibiting additives, they are claimed for use predominantly in polymers containing halogen. The effectiveness of metal compounds as smoke-suppressant additives is derived from their ability to react with degrading halogenated polymers and become involved in the degradation sequences. Many of the other additives show similar potential for involvement in degradation of their respective host polymer systems. Most additives have an advantage over fillers in that their activity is catalytic in nature, not dependent on mass–physical effects, which allows them to be used at much lower concentrations than fillers. As a result, effects on polymer physical properties can be minimized. Although most of the examples listed above do give significant reductions in smoke generation, not all are equally desirable, for reasons of economics, pigmentation or coloration, oxidant characteristics, processibility, toxicity, effects on thermal stability, electrical properties, etc. Careful attention must be given to these and other considerations in selecting an additive for use in a particular polymer system.

3.2.1.3. Surface Treatments. Only a few examples or claims to the use of surface treatments for reducing smoke are known. This may be partly due to the fact that smoke reduction is an unrecognized attendant benefit in surface treatments for flame resistance. The role a surface coating would be expected to play is one of insulation, by removing the substrate polymer from involvement in combustion and thus limiting its smoke contribution. A prime example would be an intumescent paint, which chars and expands upon exposure to heat, forming a layer which insulates the polymer surface below.[145] Another example, in which smoke inhibition is specifically claimed, is the electroless plating of polyblends with a metal such as Ni, Cu, or Cr.[146]

An especially interesting approach is the application of a "flame-barrier" interliner to an otherwise flammable, smoking polymeric article. For example, a polyurethane or polychloroprene foam seat cushion can be wrapped with a thin sheath of a sacrificial foam which is highly flame and smoke retarded. On exposure to fire, the outer wrap forms an insulating layer around the untreated core. DuPont's Vonar® interliner, a polychloroprene latex foam, heavily loaded with flame-retardant fillers and additives, is offered for this purpose,[147] as is Kimbar,® a flame-resistant nonwoven fabric developed by the Kimberley-Clark Co.

3.2.1.4. Polymer Modifications. Most of the treatments described above use variations in compounding or coating to change the smoke characteristics of a given polymer. It is also possible to improve smoking behavior by performing fundamental changes in the structure or composition of the polymer. In principle, this might be accomplished by selecting alternative polymers, or modifying polymers to obtain (a) clean-burning pyrolyzates, (b) increased thermal stability, (c) increased charring, and/or (d) slower rates of burning. Some examples of these approaches are summarized below.

a. *Polymers with Clean-Burning Pyrolyzates.* From Table 4–6 and Fig. 1, polymers with greater and lesser tendencies to form smoke can be distinguished. For a given application, if only smoke properties were critical, a polymer which burns cleanly (such as an acrylate) could be selected over one which is smoky (such as a polystyrene). Matters are rarely that simple and factors such as flammability, physical properties, and processing requirements and cost must also be considered.

Several modifications of polymer compositions to alter the pyrolyzates have been reported, including the copolymerization of methacrylate monomers with α-methylstyrene to obtain a thermoplastic with less smoke than

TABLE 22
Reversion Temperatures of Structures Commonly Found in Polyurethanes[a]

Structure	Name	Approximate temperature of reversion (°C)
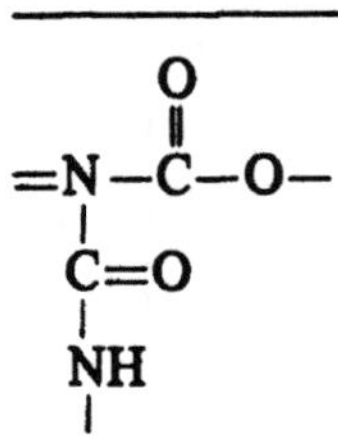	Allophanate	120–125
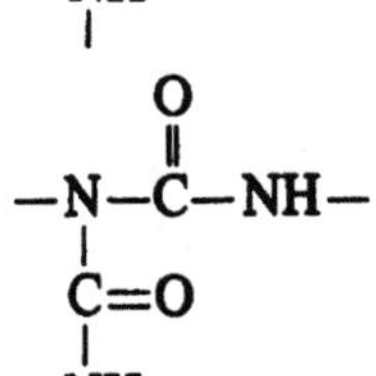	Biuret	120–125
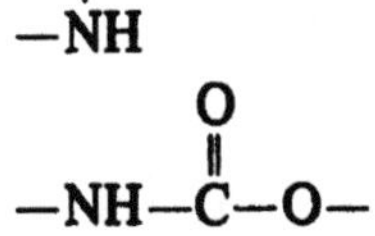	Urethane	150
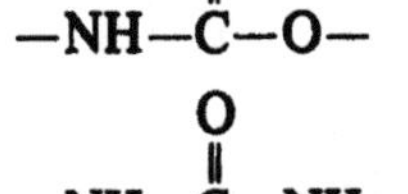	Urea	160
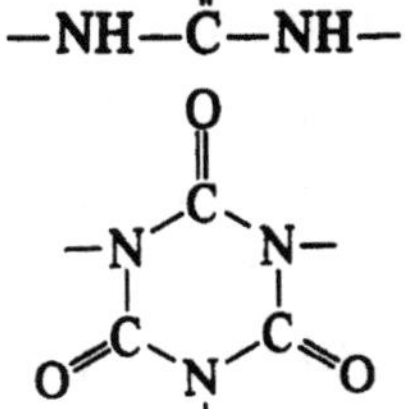	Isocyanurate	>270

[a]Reference 150.

poly(α-methyl styrene),[148] and the use of methyl methacrylate and similar monomers to cross-link polyester thermosets.[149]

b. *Polymers with Increased Thermal Stability*. Many examples of this approach are provided by the use of polyisocyanurate-based foams for insulation materials. Table 22 contains a list of structures typically found in polyurethanes, as well as the relative thermal stability of these structures. Isocyanurate linkages, with reversion temperatures on the order of 270° C and greater, are more stable than urethane linkages, which revert at about 150° C.[150] The greater thermal stability of the isocyanurate structure is believed to contribute to lower flammability and smoke generation in polyisocyanurates, as compared to polyurethanes.[151–154]

Some other polymers which show high thermal stability have aromatic or heterocyclic linkages comprising the backbone. Repeat-unit structures for several of these are shown below:

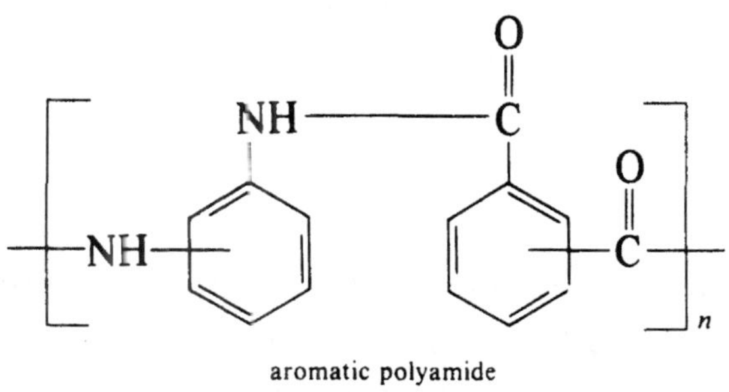

aromatic polyamide

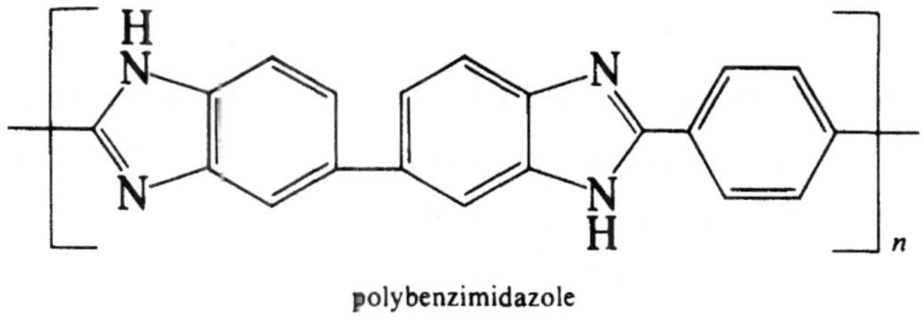

polybenzimidazole

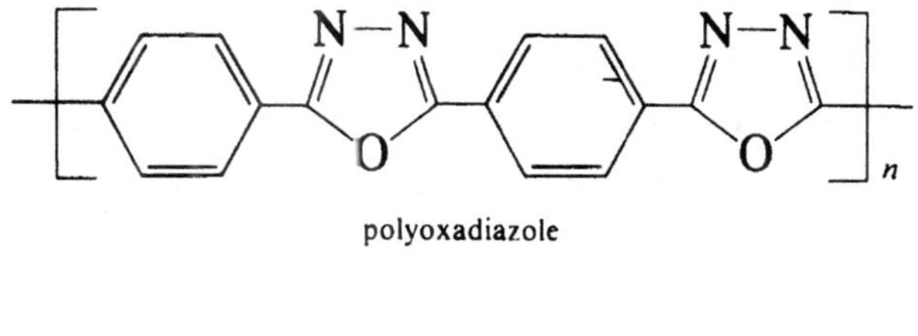

polyoxadiazole

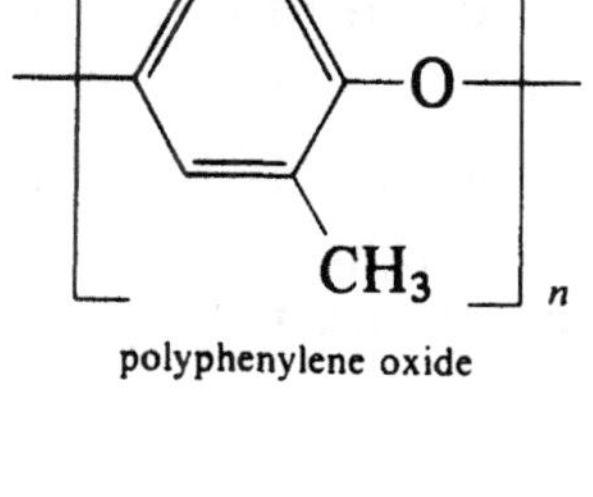

polyphenylene oxide

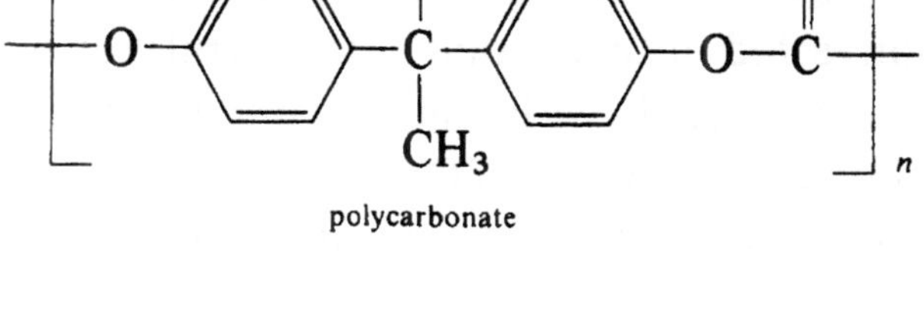

polycarbonate

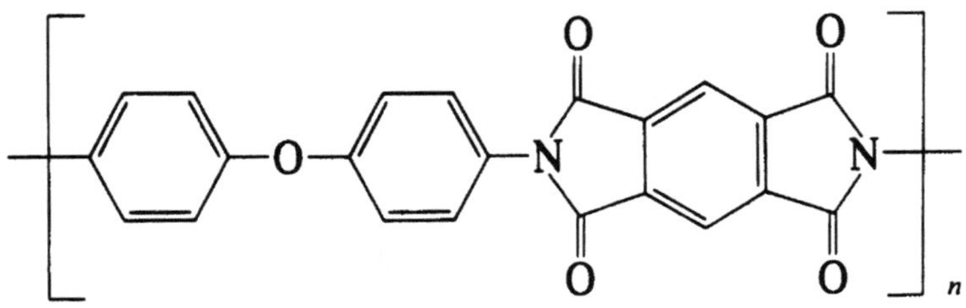

poly(oxydiphenyl pyromellitimide)
(polyimide)

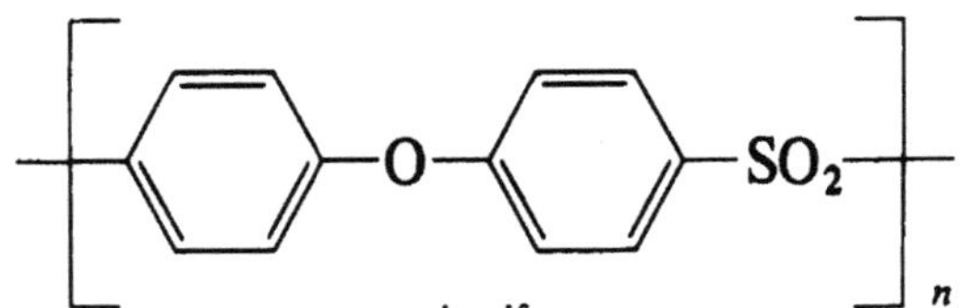

polysulfone

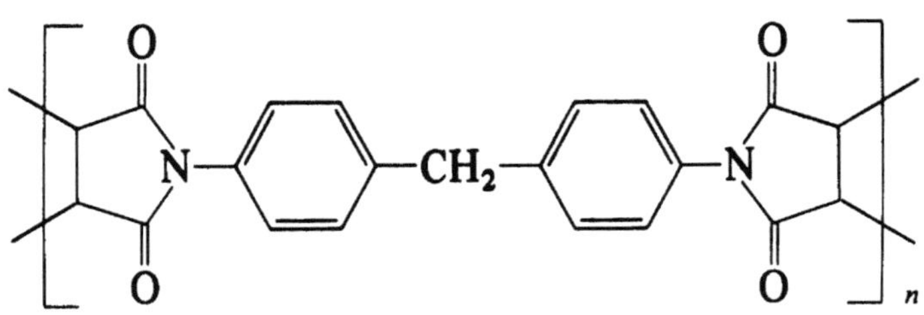

bis(maleimide)

Polyimides and polybenzimidazole, for example, decompose only at high temperatures with little off-gassing, producing large amounts of solid phase char and little smoke. Bismaleimides are useful in resin compositions, giving materials with low flammability and low smoke generation.[155] The low smoke generation of these and other polycarbonates, polyphenylene oxides, and polysulfones has prompted their investigation for various aircraft interior applications.[156–158] Drawbacks of many of these polymers include difficult processibility, coloration, and high cost.

c. *Polymers Which Form Char*. Polymers with aromatic backbones, such as those described in the preceding paragraph, generally form large amounts of solid char when burnt. This, in addition to their thermal stability, undoubtedly contributes to their low-smoke properties, since the release of aromatic pyrolyzates into the dispersed phase is minimized.

By postchlorination of PVC, a thermoplastic chlorinated polyvinyl chloride (CPVC) is obtained. This material may contain anywhere from 61% to 72% by weight chlorine, and is suggested for use in plastic pipe, etc. CPVC exhibits marked improvements in flammability (OI $\sim$ 60) and smoke generation [D_m(F) $\sim$ 60] over unmodified PVC [OI $\sim$ 43, D_m(F) $\sim$ 400].[159] Studies of the pyrolysis of CPVC have shown that about 25% of the initial polymer is converted into a solid char, while only about 6% is volatilized as a combustion fuel, with the remainder appearing as gaseous HCl.[160] Polyvinylidene chloride, PVDC, which contains 73% by weight chlorine, also forms more char and generates less smoke than PVC. PVDC has been substituted for PVC in a low-smoke flexible vinyl formulation.[161] The smoke properties of PVDC and CPVC have been attributed to both the charring and the nature of the pyrolyzates generated.[162,163]

Poly(organophosphazenes), **VI**, also called "phosphonitrilics," are a rather unique class of polymers which have backbones comprised of alternating phosphorus and nitrogen atoms. Flame-resistant and low-smoke properties have been ascribed to them, and they show interesting potential for

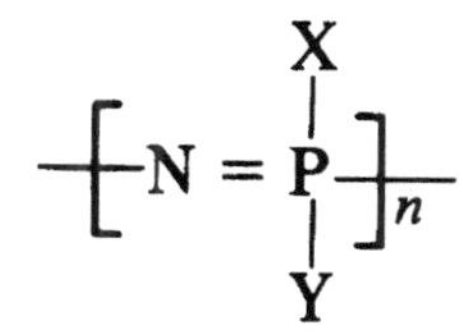

VI:
polyorganophosphazene
(X,Y = OR, OAr, or NR_2)

many applications.[164] By varying the nature of the X and Y substituents, a broad range of physical properties are attainable, and the thermal stability, flammability, and amount of charred residue formed on burning vary as well

TABLE 23
Oxygen Index and NBS Smoke Density Results for Filled and Unfilled Aryloxyphosphazene Films and Foams[a]

Polymer	Filler	(phr)	OI	NBS smoke density[b]: Flaming D_{mc}	Flaming D_m/g	Nonflaming D_{mc}	Nonflaming D_m/g
Polyethylene	None	–[c]	17	150	–	470	–
PVC	None	–[c]	44	530	–	490	–
Silicone rubber	None	–[c]	26	380	–	240	–
$(-N{=}P(OC_6H_5)(OC_6H_5))_n$	None	–[c]	34	320	50	200	36
	$Al(OH)_3$	(50)[c]	38	210	49	260	45
	$CaCO_3$	(50)[c]	40	300	63	140	29
	SiO_2	(50)[c]	30	180	47	140	39
$(-N{=}P(OC_6H_4OCH_3)(OC_6H_4OCH_3))_n$	None	–[c]	26	120	27	23	9
	$Al(OH)_3$	–[c]	44	73	18	24	6
	$CaCO_3$	(50)[c]	30	40	9	17	3
	SiO_2	(50)[c]	30	81	22	45	11
$[-N{=}P(OC_6H_4OCH_3)(OC_6H_4(sec\text{-}C_4H_9))]_n$	None	–[c]	25	180	32	–	–
	$Al(OH)_3$	(50)[d]	30	80	20	–	–
	$Al(OH)_3$	(150)[d]	48	85	8	–	–

[a]References 171 and 172.
[b]D_{mc} rounded to nearest 10.
[c]Film.
[d]Peroxide-cured foam, Ref. 172.

(1–52% char residue at 600° C, nitrogen atmosphere).[165] Although the smoke properties of elastomeric poly(aryloxyphosphazene) esters alone are not exceptional,[166] the polymers are very adaptable to the use of fillers, which improve smoke properties markedly. Rubber foams based on poly(aryloxyphosphazenes), wherein a high degree of char formation occurs, have been reported to have excellent flame resistance and low-smoke properties.[167,168] Smoke generation tendencies of these materials can be further optimized for foam and wire covering applications through the use of fillers[169] and additives,[170] and by design of the polymer through choice of ester substituents.[165,171,172] Table 23 indicates the effect on smoke generation caused by variations in the ester structure. Mechanistic reasons for improved smoke generation in the (alkoxy)aryloxy derivatives have not been demonstrated, although the pyrolytic fragments expected would be more oxygenated and perhaps cleaner burning. The poly(aryloxyphosphazenes) are of continuing interest since flammability and smoke generation properties are improved in the absence of halogens, and a low order of toxicity of combustion products has been reported.[173]

d. *Miscellaneous Polymer Modifications.* Numerous other methods of modifying polymer structures to improve smoke generation are possible. In polyurethanes, for example, hydrophilic polyols can be used to enhance the retention of moisture in the foam, retarding both flammability and smoke generation when hydrated fillers are employed.[174] Polyols based on dibromobutenediol[175], **VII**, or trichlorobutylene oxide[176–178], **VIII**, are also reported to give polyurethanes with less smoke than conventional systems,

VII:

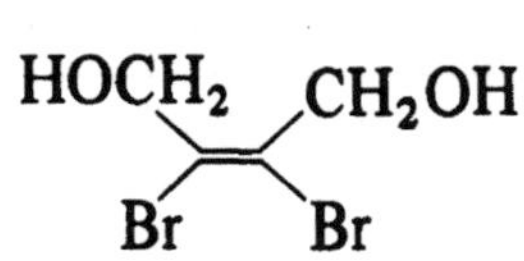

VIII:

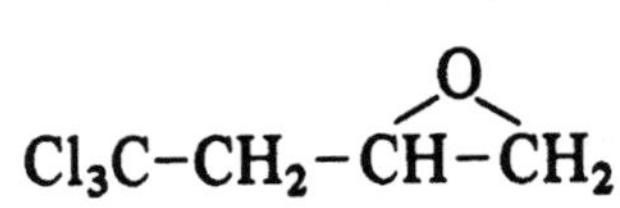

presumably because of their reduced flammability. The addition of diethanolamine[179] or dimethylol urea[180] to the polyol also reportedly gives polyurethane foams with improved smoke generation properties.

The intensity and rate of smoke generation from styrene–butadiene thermoset resins with high vinyl content were reduced by the complexation of a portion of the styrene units with chromium carbonyl, e.g.,[181]

IX:

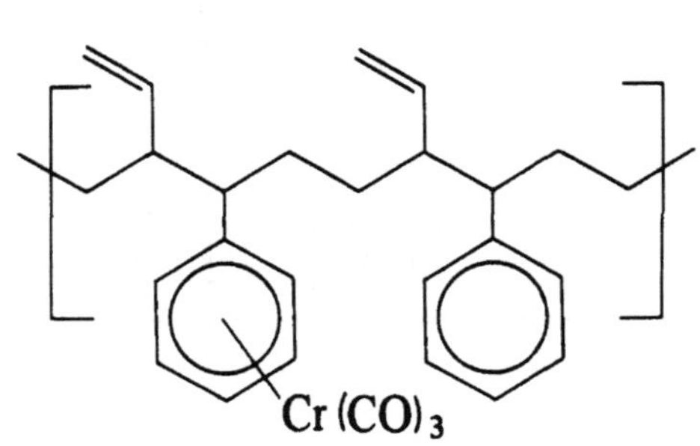

Evidence of smoke inhibition by both gas and solid phase processes was found.[182]

3.2.2. Methods for Smoke Reduction Classified by Type of Polymer

A summary of smoke-suppressant treatments used in polymer applications is given in Table 24. The listing is according to the type of polymer treated. Because of the number of references and the fact that most data are reported in patents, discussion of individual approaches will not be given. Readers may refer to the original citations and Ref. 72 for more information.

TABLE 24
Methods of Smoke Reduction Reported for Various Polymers

Treatment	Details	Refs.
A. PVC and halogenated polymers		
1. Fillers		
$Al(OH)_3$		75, 142
Dawsonite	Reinforcement properties reported	142
MgO	Chlorosulfonated polyethylene	183
Basic magnesium carbonate	Phosphate-plasticized PVC; PVC–butadiene–acrylonitrile blends	141, 142, 184
$MgCO_3$, MgO, $Mg(OH)_2$ hydrotalcite	PVC and CPE blends with ABS	185
Alkaline earths, B compounds	Flexible PVC wire insulation with nonphosphate plasticizer	186
$Mg_2B_2O_4$, $CaB_6O_{11} \cdot 5H_2O$	~60% smoke reduction claimed	187
2. Additives		
Metal compounds:		
Ferrocene	Optimum at less than 0.5% by weight, improved flame resistance and smoke	47, 91, 92
DFR 100®, DFR 121®	Proprietary ferrocene-based systems, recommended for polyurethane and flexible vinyl	92, 96, 97
Fe powder	Used alone or in combination with other metal oxides	98, 102, 103
Fe oxide or hydroxide	2.5 phr in vinyl chloride copolymer foams	99
FeS	~50% reduction of smoke used 1–10 phr	188
Fe^{+2} trimellitate	0.5–2 phr used, Sn stabilizer may interfere	124
MoO_3	Replacement for Sb_2O_3	98, 102–105
Ammonium molybdates	Similar to MoO_3	104
MoO_3	Can be used alone or in mixtures	189
Mo compounds	Improve smoke and flammability	190–192

TABLE 24 (Continued)

Treatment	Details	Refs.
Melamine molybdate	Used alone or in mixtures	193
ZnO, Zn compounds	Usually used at less than 1 phr	106, 194–198
Ongard® I, II	Proprietary Mg-Zn complex, can replace Sb_2O_3	107–109, 199
$MgO \cdot ZnSO_4$	MgO complex of Zn salts, sulfates, fluoroborates, fluorosilicates	200
$ZnO–CaCO_3$	EVC copolymer, 8 phr ZnO, 80 phr $CaCO_3$	201
Zn borate	VC-vinylphosphonate copolymer	202
CuCN, CuSCN	Reduced D_m as much as 70%, also increased OI	203
Cu_2S, CuS, Cu oxalate	Used at 1–10 phr	188, 204
Thermoguard® S-711	Sb_2O_3 modified with Na antimonate, MgO	205, 206
Oncor® RAZ	Modified Sb_2O_3, containing Zn	207
Ni or Co compound	$Ni(acac)_2$ + $(BrCH_2CHClCH_2O)_3PO$	208
Cr_2S_3 or AlB_{12}	10 phr in PVC	209
Si_3N_4	Used alone or with MoS_2	189
Sn oleate, $Cr(acac)_3$, Ni hydroxyquinone	Chlorotrifluoroethylene–ethylene copolymer	210
Sn compounds	0.1–50 phr loading used in flexible	211
Mn borate	20–40% smoke reduction in flexible using 1–10 phr	212
Fe^{+2}, Zn^{+2}, alkali metal and alkaline earth metal borates	Variable effects in PVC and PVDC	213
V_2O_5, $V^{III}(acac)_3$	At 2% by weight V, in flexible or rigid	214
Zn, Co, Mn, Fe dicarboxylates	Optionally used with $Al(OH)_3$	112, 215
Metal salts of mercaptobenzothiazole	Various transition metal cations	216
Metal salts of phosphinic acids	Example: $CuP(O)(C_6H_5)_2$	217
Mixed-metal systems:		
$K_2Zn_3(Fe(CN)_6)_2$ and similar compounds	50% reduction of D_m, increased OI	113
$KZn\,Cu(CN)_4$	0.5–5 phr decreased D_m by over 50%	114
$Zn–MoO_3$ or $ZnMoO_4$	40–50% smoke reduction with 1.5 phr $ZnMoO_4$	117–119
Zn ammonium polymolybdate	Used in both flexible and rigid	218
Zn zirconate, titanate, or tungstate	Rigid or plasticized PVC	115, 116
Zn stannate	Used in PVC–ABS blends	219
ZnO, Zn borate–$Al(OH)_3$	Strong smoke reductions, increased OI	110, 111
Bi subcarbonate–$Al(OH)_3$	Synergistic, especially in rigid PVC	220

TABLE 24 (Continued)

Treatment	Details	Refs.
Various Ni compounds + compounds of Al, Bi, Cu, Fe, Mo, Sb, V, or Zn	Synergistic combinations; many can be used individually	120–122, 221, 222
Mo salts + compounds of Co, Fe, Ni, W, Zn, borates, silicates, or silica gel	Synergistic pairs	223
Mo salts + CuO, Cu_2O, or Cu oxalate	Synergistic pairs	224, 225
Undisclosed PVC compound	Condensed-phase additives	226
Plasticizers, etc.:		
Dioctyl azelate, ethylhexyl oxalate	Used for low-smoke wire insulation	139, 140
Haloalkyl phthalates, trimellitates, benzenepolycarboxylates	In combination with active fillers, also used in ABS, PS, PE, polyurethane	227
Tricresyl phosphate + MgO	CPE–Chlorosulfonated PE blend	228
Halogenated polyphosphates	Used with hydrated metal oxides	229
Aryl, alkyl phosphates	Used with active fillers	141
C_1–C_8 borate ester	Used with phosphate plasticizer	136
3. Coatings		
Intumescent paints		230
4. Polymer substitution or modification		
Polymer substitution	PVDC with epoxy plasticizer	161
Polymer modification	Chlorinated PVC	162, 163
B. Urethanes and related polymers		
1. Fillers		
$Al(OH)_3$	Rigid foams, used with phosphonate	231
$Al(OH)_3$, silica	Used with melamine, optional boric acid	174, 213
Powdered cellulose	Used at 70% by weight in rigid foam, increase char, reduces smoke	144
2. Additives		
Ferrocene	Claimed with PVC	91
Mn trimellitate	Used with F.R. additives	232
Solid dicarboxylic and polycarboxylic acids	In polyurethanes and polyisocyanurates	124, 126, 127, 233, 234
Aromatic dicarboxylic acid derivaties	Anhydrides, acid chlorides	128
KBF_4, NH_4BF_4	Isocyanurate foams	131
Sulfur, ethylene thiocarbonate, butyl polysulfide	Rigid foams	132, 133

TABLE 24 (Continued)

Treatment	Details	Refs.
Borax	Foams containing polyphosphates and haloalkylphosphate esters	137
Melamine	Also used with boric acid, fillers	174, 213
Organosilicon compounds	Cyclic siloxanes, silicates, silicone surfactants in polyisocyanurates	143, 235–237
Aromatic alcohols, aldehydes	Benzyl, furfuryl alcohols, benzaldehyde, furfuraldehyde in polyisocyanurates	130, 238
3. Surface coatings		
Flame barrier wrap	Polychloroprene foam with high filler loading, wrapped around polyurethane and other foams	147
4. Polymer substitution or structural modifications		
Dibromobutenediol-based polyol	Mixed with diphosphonate	175
Trichlorobutylene oxide polyol	For rigid foams	176, 177, 178
Hydrophilic polyols	Used with active filler, etc.	174
Phthalate and trimellitate-based polyester polyols	Used for polyurethanes and polyisocyanurates	239, 240
HMDI, CHMDI	Substituted for TDI	241
Diethanolamine	With mixed TDI, DPMDI–polyol	242
Diethanolamine	Used with polyol, saturated with CO_2, then treated with TDI	179, 243
Methylolurea	Added to polyol component	180
Diethylene glycol–glycerin polyol	50% Trimerized isocyanate, containing 20% polyacrylonitrile	152
Trimerized structures	Equal parts polyol and trimerized isocyanate	151, 153
C. Polyesters and related resins		
1. Fillers		
$Al(OH)_3$	Minimizes smoking in various resin composites	77, 149, 244
Barytes	Used in halogenated polyesters	245
SiO_2	Used with boric acid, borax in glycol–maleic anhydride resin	135
$Mg(OH)_2$	Most useful in non-halogenated resin	85, 245
Al_2O_3, SiO_2 fibers	Fill voids in phenolic, polyurethane, or polystyrene foams	246
Graphite fibrils	Various resin composites	247
2. Additives		
Fe^{+2} oxalate	Halogenated polyester resin	100
Fe compound + Cu or Sb compound	Increased carbonization, reduced flammability and smoke in halogenated resin	248, 249

TABLE 24 (Continued)

Treatment	Details	Refs.
BaZn borate, CaMg phosphate	Chlorinated polyester	250
MoO_3	In polyester based on or containing halogen	192, 251
Fumaric, maleic, succinic acids	Halogenated polyesters	252
Ammonium sulfamate	Phenolic foams	134
3. Polymer substitutions and structural modifications		
Acrylates	Incorporated as cross-linking monomer	149
Vinyl phosphates	In maleic anhydride polyester	253
Furfural, paraformaldehyde	Ingredients in flexible glass-reinforced resorcinol, phenolic resins	254
Undisclosed structure	"Low-smoke" Narmco® fiber-reinforced prepregs	255
D. Polymers of styrene and/or butadiene, and rubber		
1. Fillers		
$Al(OH)_3$, dawsonite, clays	SBR, NR, polystyrene, ABS, EPDM	75, 76, 82, 84
Intumescent filler	With $Al(OH)_3$ in SBR and other elastomers	256
$Mg(OH)_2$	SBR, NR, chlorinated latex rubber, EPDM	80, 83, 84
Basic Mg carbonate	BAN foams with PVC	184
$MgO \cdot xMgCl_2 \cdot yH_2O$	SBR, polychloroprene	257, 258
2. Additives		
Ferrocene	30-60% reduction of smoke during incineration of polystyrene	259
Fe_2O_3	BAN-PVC foams filled with basic Mg carbonate	184
Mo oxides and salts	ABS, butadiene rubber containing halogens	260, 261
Phthalocyanine complexes of Fe, Cu, Mn, Co, VO	In PS, ABS–use compatible with halogens	262
Fe, Mn, Cr hydroxyquinolates	1-2.5% as metal reported effective in PS, ABS, SAN, SBR, etc.	263
Mercaptobenzothiazole salts	Various transition metal cations	216
$(C_6H_5)_4Pb$	Smoke mass reduced by 50% with 6% by weight level in ABS	264
Zn borate	Polychloroprene rubber	257
Melamine-formaldehyde resin	Polychloroprene rubber, post-treatment of sponge	257, 265
3. Surface coatings		
Intumescent paint	Coated PS beads expanded to foam	266
Metal plating	Electroless plating of Ni, Cu or Cr to polychloroprene-SAN polyblends	146

TABLE 24 (Continued)

Treatment	Details	Refs.
4. Polymer substitution or structural modification		
C_1–C_8 methacrylates	Copolymerized with α-methylstyrene	148
Methacrylate monomers	Graft copolymerized with styrene, acrylonitrile from polychloroprene	267
$Cr(CO)_3 \cdot$ SBR complex	High vinyl SBR molding compound	181, 182
E. Polyolefins		
1. Fillers		
$Al(OH)_3$	Polypropylene and polyethylene	75
$CaCO_3$	Polypropylene, polyethylene with organometallic additives	268, 269
2. Additives		
Ferrocene, nickelocene, $Al(acac)_3$	Polypropylene, polyethylene with $CaCO_3$ filler	268, 269
Organophosphoranes	Polypropylene used with ammonium polyphosphate and TiO_2	270
ZrO_2	Polyethylene containing chlorinated hydrocarbon, optional $Al(OH)_3$	271
Zn compound, PVC	PVA fibers impregnated with mixture	272
$ZnCl_2$	PVAC adhesive for vinyl wall covering	273
F. Miscellaneous Polymers		
1. Fillers		
$Al(OH)_3$	Various polymers	36, 75 274–277
$CaCO_3$, ammonium polyphosphate	Smokeless plastic matchsticks	278
$Al(OH)_3$-titanate-treated	Polyphosphazenes	279
$Al(OH)_3$, basic $MgCO_3$, $Mg(OH)_2$	Polyphosphazenes, including plasticized, useful for foams, wire coverings, etc.	169
2. Additives		
Borax	Rayon wall covering	280
$Na_2Cr_2O_7$, H_3BO_3, $(NH_4)_2SO_4$, borax, glycerin, melamine–formaldehyde resin, ammonium polyphosphate	Plywood	281–283
Trimesic acid	Poly(aryloxyphosphazenes)	170
KCN	Polyacrylonitrile, especially with acrylic acid monomers	284

TABLE 24 (Continued)

Treatment	Details	Refs.
Iron salts	Various polymers, paints, etc.	285
3. Coatings, finishes, etc.		
Silicates	Finish for woods	286
Hexafluorozirconates, titanates	Wool finish	287
Carboxylic acids	Smoke density of wool affected by pH	288
4. Polymer substitution or structural modification		
Polysilicones	Substituted in cable jacket, aircraft seat cushions	289, 290
Aromatic polyimides	Coating on fabric–honeycomb laminate	291
Bismaleimides	Carbon composites with high thermal stability	156
Polycarbonates, polyphenylene-oxides, polysulfones, poly-silicones	Used in aircraft interiors	157, 158
Polyphosphazenes	Blended with polysilicones	292
Polyphosphazenes	Optimization of smoke and flame properties with alkoxyaryloxy substituents	171, 172 293, 294
Polyphosphazenes	Blended with BAN, EPR, urethane	295
Polyphosphazenes	Low-smoke, flame-resistant rubber foams	167, 169, 296

4. Summary and Conclusions

The references cited in the previous discussions and listed in Table 24 are not exhaustive, but are meant to be an accurate representation of the state-of-the-art of smoke inhibition for polymers. The effectiveness, practicality, and economics of these approaches vary widely, but the composite list does serve to illustrate several factors which favor reduced smoke generation, whether achieved by polymer selection, the use of additives or fillers, or polymer modifications:

a. *Low aromatic or conjugated polyene content of fuels* Paraffinic and oxygenated polymers generally show lower smoke generation tendencies than other systems.

b. *Increased polymer stability.* Achieved through inherent structural features or modifications which limit decomposition to form smoky combustion fuels.

c. *Char formation and the retention of combustible substrates in the solid phase.* Polymers which display this behavior include chlorinated PVC and polymers with backbone aromatic groups; hydrated fillers and some of the metal-containing additives promote char in certain polymers.

d. *Reduced burning rates to allow fuel/oxidant ratios favoring clean burning.* Achieved with some flame retardants, provided that adverse factors, such as increased aromatic content of fuels, are not also promoted. Hydrated fillers may contribute to this in some polymers.

e. *Enhanced oxidation of combustion products, including soot.* Various metal-containing additives may function by catalyzing the destruction of soot.

f. *Alteration of the light transmission characteristics of smoke particles.* Although not mentioned in the discussion, variations in particle size ranges or absorptivity of smoke particles can affect the measured smoke intensity.

The most desirable of these approaches would appear to be those which minimize the generation of combustible species, such as through increased polymer stability or char formation. This not only contributes to the reduction of visible smoke, but also tends to reduce flammability and limit the volume of combustion products formed. Considerable improvements in smoke behavior have been achieved for certain polymers, notably vinyls, by taking advantage of synergisms between additives. Further developments using synergisms might be anticipated, particularly where additive synergisms are combined with a structural modification.

The ultimate utility and commercial practicality of any of the methods for smoke reduction described here still remain to be demonstrated. This will first require validation of currently used smoke tests with full-scale, real-life fires, and then careful evaluation of the methods in end-use systems.

Acknowledgments

This survey is based on lectures presented by the author in short courses of the Plastics Institute of America and in the Polymer Conference Series of the University of Detroit, parts of which were published as a review.[72] Portions of the review article are reprinted in this text with permission of the copyright holder, Technomic Publishing Co., Inc., Westport, Connecticut. Portions of the indicated tables and figures are reprinted with permission of the copyright owners listed below:

American Chemical Society, Washington, D.C., Table 1.

Technomic Publishing Co., Westport, Connecticut, Tables 2, 3, 5, 6, 8, 9, 11, 13, 14, 18, 23, 24, Fig. 1.

Society of Plastics Engineers, Greenwich, Connecticut, Table 4.

Rubber Division, American Chemical Society, Akron, Ohio, Table 10.

John Wiley & Sons, Inc., New York, New York, Table 12.

National Fire Protection Association, Boston, Massachusetts, Table 7.

The Plastics and Rubber Institute, London, England, Scheme I.

The support and permission of the Firestone Tire & Rubber Co., especially Dr. Alan G. Altenau, to publish this survey is greatly appreciated.

No approval to practice any invention discussed herein without proper approval and licensing is hereby intended.

NOTE: The indices of flammability and smoke generation described herein are relative results based on laboratory tests and are not intended to reflect hazards presented by any material under actual fire conditions. The terms *smoke-suppressant*, *smoke-inhibited*, *smoke-retardant*, or *retarded*, etc. are used in reference to the behavior of materials in laboratory tests and are not intended to reflect behavior under actual fire conditions.

5. *References*

1. National Commission on Fire Prevention and Control, *American Burning*, U.S. Government Printing Office, Washington, D.C. (1973).
2. R. B. Engdahl, in *Encyclopedia of Chemical Technology*, 2nd ed., Vol. 18, pp. 400–415, Kirk-Othmer Publishers, New York (1969).
3. C. J. Hilado, *J. Fire Flammability* **1**, 217 (1970).
4. H. Th. Hoffmann and H. Oettel, *Mod. Plast.* **46**(10), 94 (1969).
5. N. K. Raman and K. A. Scott, *RAPRA Tech. Rev.* **52** (June, 1972).
6. B. W. Mapperly and P. R. Sewell, *Eur. Polym. J.* **9**, 1255 (1973).
7. M. Chaigneau, *C. R. Acad Sci. Ser. C* **278**, 109 (1974).
8. J. R. Gaskill, *SPE J.* **28**, 43 (1972).
9. R. E. Reinke and C. R. Reinhardt, *Mod. Plast.* **50**(2), 94 (1973).
10. National Materials Advisory Board, *Fire Safety Aspects of Polymeric Materials, Vol. 3, Smoke and Toxicity (Combustion Toxicology of Polymers)*, Technomic Publishing Co., Westport, Connecticut (1978).
11. M. M. Birky, *Polym. Prepr. Am. Chem. Soc. Div. Polym. Chem.* **14**(2), 1011 (1973).
12. J. J. Kracklauer and C. J. Sparkes, SPE Regional Technical Conference, Boston, Massachusetts, November 1976.
13. P. A. Croce, Factory Mutual Research Report, No. 21011.4, RC75-T-31, 1975.
14. G. Heskestad, Escape Potentials from Apartments Protected by Fire Detectors in Highrise Buildings, NTIS Report No. PB-234014 (1974), p. 7.
15. R. W. Bukowski, T. E. Waterman, and W. J. Christian, Detector Sensitivity and Siting Requirements for Dwellings, NTIS Report No. PB 247483 (1975), pp. D1–D8.
16. M. Faraday, *The Chemical History of a Candle*, Thomas Y. Crowell Co., New York (1957).
17. a. J. H. Knox, R. G. W. Norrish, and G. Porter, *J. Chem. Soc.*, 1447 (1952); b. G. Porter, *Combustion Researches and Reviews*, p. 108, Butterworth Publishing Co., London (1955).
18. a. U. Bonne, K. H. Homann, and H. G. Wagner, *Symp. (Int.) Combust.* **10**, 503 (1965); b. K. H. Homann and H. G. Wagner, *Proc. R. Soc. London Ser. A* **307**, 141 (1968).
19. E. E. Tompkins and R. Long, *Symp. (Int.) Combust.* **12**, 625 (1969).
20. J. Kern and G. Spengler, *Erdöl Kohle Erdgas Petrochemie*, 813 (1970).
21. K. H. Homann and H. G. Wagner, *Sympos. (Int.) Combust.* **11**, 371 (1967).
22. P. C. Warren, in *Polymer Stabilization* (W. L. Hawkins, ed.), pp. 326–330, Wiley-Interscience, New York (1972).
23. A. M. Calcraft, R. J. S. Green, and T. S. McRoberts, *Plast. Polym.* **42**(161), 200 (1974).
24. A. Thomas, *Combust. Flame* **6**, 46 (1962).
25. R. A. Hunt, Jr., *Ind. Eng. Chem.* **45**, 602 (1953).
26. P. G. Edgerley and K. Pettett, *Fire Mater.* **2**, 11 (1978).

27. C. P. Bankston, E. A. Powell, R. A. Casanova, and B. T. Zinn, *J. Fire Flammability* **8**, 395 (1977).
28. C. P. Bankston, R. A. Casanova, E. A. Powell, and B. T. Zinn, *J. Fire Flammability* **7**, 165 (1976).
29. R. W. Welker and J. P. Wagner, *J. Fire Flammability* **8**, 26 (1977).
30. B. T. Zinn, E. A. Powell, R. A. Casanova, and C. P. Bankston, *Fire Res.* **1**, 23 (1977).
31. I. A. Benjamin, F. Fung, and L. Roth, NBS IR 77-1209, National Bureau of Standards, Washington, D.C. (1977).
32. J. D. Seader and W. P. Chien, *J. Fire Flammability* **5**, 151 (1974).
33. K. Maries, *Fire Mater.* **2**, 2 (1978).
34. S. K. Brauman, N. Fishman, A. S. Brolly, and D. L. Chamberlain, *J. Fire Flammability* **6**, 41 (1975).
35. S. K. Brauman, *J. Polym. Sci. Polym. Chem. Ed.* **15**, 1507 (1977).
36. G. L. Nelson, *J. Fire Flammability* **5**, 125 (1974).
37. A. J. Steiner, Underwriters Laboratory Research Bulletin No. 32, (1944); ASTM E-84, ASTM Std. **14**, 472 (1972).
38. C. C. Peters and H. W. Eickner, ASTM STP 344, 18 (1962); ASTM E-286, ASTM Std. **14**, 586 (1972).
39. F. J. Rarig and A. J. Bartosic, ASTM STP 422, 106 (1967); ASTM D-2843, ASTM Std. **27**, 712 (1972).
40. D. Gross, J. J. Loftus, and A. F. Robertson, ASTM STP 422, 166 (1967).
41. A. F. Robertson, ASTM STP 344, 33 (1972); ASTM E-162, ASTM Std. **14**, 558 (1972).
42. R. E. Cass, *J. Cell. Plast.* **3**, 41 (1967).
43. J. DiPietro, H. Barda, and H. Stepniczka, *Text. Chem. Colorist* **3**, 40 (1971).
44. J. R. Gaskill, *J. Fire Flammability* **4**, 279 (1973).
45. R. W. Mickelson, in Polymer Conference Series, University of Utah, Salt Lake City, Utah, July 1972.
46. L. P. Parts, Polymer Conference Series, University of Utah, Salt Lake City, Utah, July 1972.
47. J. J. Kracklauer and C. J. Sparkes, *Plast. Eng.* **11**(6), 57 (1974).
48. E. E. Smith, *Fire Technol.* **8**, 237 (1972).
49. R. Jacob, in Proceedings of International Conference on Non-Inflammability of Plastics, Wood and Textiles, Strbske Pleso, Czechoslovakia, September 1974; proposed French Standard NF-T51-073, 1977.
50. L. G. Imhof and K. C. Steuben, *Polym. Eng. Sci.* **13**, 152 (1973).
51. J. R. Gaskill and C. R. Veith, *Am. Chem. Soc. Div. Org. Coat. Plast.* **1**, 282 (April 1968); *Fire Technol.* **4**(3), 185 (1968).
52. C. J. Sparkes, J. J. Kracklauer, and R. E. Legg, in 26th International Wire and Cable Symposium, Cherry Hill, New Jersey, November 1977.
53. J. B. Fang, *NBS Tech. Note*, No. 879 (June 1975).
54. C. J. Hilado, *Flammability Handbook for Plastics*, pp. 42 and 43, Technomic Publishing Co., Westport, Connecticut (1969).
55. J. D. Seader and S. S. Ou, *Fire Res.* **1**, 3 (1977).
56. L. H. Breden and M. Meisters, *J. Fire Flammability* **7**, 234 (1976).
57. L. Benisek and W. A. Phillips, *J. Fire Flammability* **8**, 516 (1977).
58. M. M. O'Mara, W. Ward, D. P. Knechtges, and R. J. Meyer, in *Flame Retardancy of Polymeric Materials* (W. C. Kuryla and A. J. Papa, eds.), Vol. 1, pp. 209–213, Marcel Dekker, New York (1973).
59. M. I. Jacobs, *J. Fire Flammability* **6**, 347 (1975).
60. A. F. Grand, *J. Fire Flammability* **7**, 216 (1976).
61. D. G. Sauers, *J. Fire Flammability* **7**, 181 (1976).
62. J. R. Gaskill, R. D. Taylor, H. W. Ford, and H. H. Miller, *J. Fire Flammability* **8**, 160 (1977).

63. W. P. Chien and J. D. Seader, *Fire Technol.* **10**, 187 (1974).
64. K. C. Salooja, *J. Inst. Fuel* **45**, 37 (1972).
65. J. W. Hastie, *J. Res. Nat. Bur. Std. Sc. A* **77A**, 733 (1973).
66. M. W. Shayeson, SAE, Paper No. 670866, October 1967.
67. K. Addecott and C. W. Nutt, *Am. Chem. Soc. Div. Pet. Chem. Prepr.* **14**, A-4, 69 (1969).
68. D. C. Turley, D. L. Brenchley, and R. R. Landolt, *J. Air Pollut. Contr. Assoc.* **23**, 783 (1973).
69. R. L. Weeks, W. L. Clinkenbeard, and J. D. Soltis, in Proceedings of the 5th World Petroleum Congress, Sect. VI, p. 381, New York (1959).
70. B. Toone, in *Combustion in Advanced Turbine Systems* (I. E. Smith, ed.), p. 271, Pergamon Press, London (1968).
71. D. R. Jenkins, *Combust. Sci. Technol.* **5**, 245 (1972).
72. D. F. Lawson and E. L. Kay, *J. Fire Retard. Chem.* **2**, 132 (1975).
73. G. Kraus, *Reinforcement of Elastomers*, Wiley-Interscience, New York (1965).
74. H. Frood and H. P. Alger, Brit. Pat. 183,922 (1921).
75. I. Sobolev and E. A. Woycheshin, *J. Fire Retard. Chem.* **1**, 13 (1974).
76. K. C. Hecker, R. E. Fruzzetti, and E. A. Sinclair, *Rubb. Chem. Technol.* **45**, 1692 (1972); abstract of paper presented at a meeting of the Rubber Division American Chemical Society, Boston, Massachusetts, April 1972.
77. P. V. Bonsignore and J. H. Manhart, SPI 29th Annual Conference, Reinforced Plastics and Composites Institute, 1974.
78. C. Eyraud, R. Goton, and M. Prethe, *C.R. Acad. Sci. Paris* **240**, 862 (1955).
79. G. V. Jackson and P. Jones, *Fire Mater.* **2**, 37 (1978).
80. D. F. Lawson, E. L. Kay, and D. T. Roberts, Jr., *Rubb. Chem. Technol.* **48**, 124 (1975).
81. D. F. Lawson, unpublished results, 1974.
82. I. Sobolev and E. A. Woychesin, *Am. Chem. Soc. Div. Org. Coat. Plast.* **36**(2), 497 (1976).
83. D. F. Lawson, D. T. Roberts, Jr., and E. L. Kay, U.S. Pat. T952009 (1976).
84. M. Moseman and J. D. Ingham, in Papers, 113th Rubber Division, American Chemical Society, Montreal (May 1978).
85. Brit. Pat. 1,080,468 (1967) (to Burns and Russell Co.).
86. J. J. Pitts, in *Flame Retardancy of Polymeric Materials* (W. C. Kuryla and A. J. Papa, eds.), Vol. 1, pp. 133–194, Marcel Dekker, New York (1973).
87. S. K. Brauman and A. S. Brolly, *J. Fire Retard. Chem.* **3**, 66 (1976).
88. S. K. Brauman, *J. Fire Retard. Chem.* **3**, 116 (1976).
89. S. K. Brauman, *J. Fire Retard. Chem.* **3**, 138 (1976).
90. J. M. Schwarcz, *J. Fire Retard. Chem.* **1**, 78 (1974).
91. J. J. Kracklauer, Ger. Pat. 2,307,387 (1973); U.S. Pat. 4,049,618 (1977) (to Syntex Corp.).
92. J. J. Kracklauer, Ger. Pat. 2,519,603 (1975) (to Syntex Corp.).
93. D. F. Lawson, *J. Appl. Polym. Sci* **20**, 2183 (1976).
94. L. Lecomte, M. Bert, A. Michel, and A. Guyot, *Am. Chem. Soc. Div. Org. Coat. Plast.* **36**(2), 732 (1976).
95. L. Lecomte, M. Bert, A. Michel, and A. Guyot, *J. Macromol. Sci. Chem.* **A11**, 1467 (1977).
96. Product Development Report 5041 (1974), and Data Bulletins Series 105,109, Arapahoe Chemicals, Inc.
97. J. J. Kracklauer, C. J. Sparkes, and E. V. O'Grady, in Proceedings, 1977 National Technical Conference, SPE, p. 129, Denver Colorado (November 1977).
98. L. C. Mitchell, U.S. Pat. 3,821,151 (1974) (to Ethyl Corp.).
99. S. Hosakawa, T. Honda, and H. Matsui, Japan. Pat. 74:77,968 (1974) (CA 82:44429u).
100. S. Oswitch and T. F. Barr, Fr. Pat. 2,236,893 (1975) (to Ferro Corp.).
101. J. E. Ramsbotton, *The Fireproofing of Textiles*, His Majesty's Stationery Office, London (1947).
102. L. C. Mitchell, U.S. Pat. 3,845,001 (1974); U.S. Pat. 3,903,028 (to Ethyl Corp.), 1975.

103. L. C. Mitchell and G. L. TerHaar, U.S. Pat. 3,870,679 (1975) (to Ethyl Corp.).
104. F. W. Moore and D. A. Church, in *Proceedings, International Symposium on Flammability and Flame Retardants*, Technomic Publishing Co., Westport, Connecticut (1976).
105a. F. W. Moore, *SPE Tech. Papers* 23, 414 (1977).
105b. W. H. Starnes and D. Edelson, *Macromolecules*, **12**, 797 (1979).
105c. R. P. Lattimer and W. J. Kroenke, *J. Appl. Polym. Sci.*, **26**, 1191 (1981).
106. J. Schwarcz, B. Engelbart, and E. L. White, U.S. Pat. 3,945,974 (1976) (to N. L. Industries, Inc.).
107. Data Sheet 175, NL Industries, Inc.; W. Malone and J. M. Schwarcz, *Plast. Eng.* **31**(7), 41 (1975).
108. E. J. Augustyn and J. M. Schwarcz, *SPE Tech. Papers* **23**, 202 (1977).
109. J. M. Schwarcz and E. J. Augustyn, *Am. Chem. Soc. Div. Org. Coat. Plast.* **39**, 560 (1978).
110. D. F. Lawson and E. L. Kay, U.S. Pat. 3,957,723 (1976) (to The Firestone Tire & Rubber Co.).
111. E. L. Kay, U.S. Pat. 3,985,706 (1976) (to The Firestone Tire & Rubber Co.).
112. D. F. Lawson, Ger. Pat. 2,651,840 (1977) (to The Firestone Tire & Rubber Co.).
113. A. W. McRowe, U.S. Pat. 3,822,234 (1974) (to B. F. Goodrich Co.).
114. A. W. McRowe, U.S. Pat. 3,862,086 (1975); U.S. Pat. 3,886,262 (1975) (to B. F. Goodrich Co.).
115. E. D. Dickens, U.S. Pat. 3,880,802 (1975) (to B. F. Goodrich Co.).
116. D. M. Kurtz, U.S. Pat. 3,901,441 (1975) (to B. F. Goodrich Co.).
117. T. Y. King, U.S. Pat. 3,900,441 (1975) (to Armstrong Cork Co.).
118. F. W. Moore and G. A. Tsigdinos, in *Proceedings, 5th International Symposium on Flammability and Flame Retardants, 1978*, Technomic Publishing Co., Westport, Connecticut (1979).
119. Sherwin-Williams Chem. Co., Technical Bulletins Nos. 346, 347 (1977); Nos. 350, 354, 356, 357 (1978).
120. E. D. Dickens, U.S. Pat. 3,933,742 (1976) (to B. F. Goodrich Co.).
121. E. D. Dickens. U.S. Pat. 3,962,163 (1976) (to B. F. Goodrich Co.).
122. E. D. Dickens, U.S. Pat. 3,983,086 (1976) (to B. F. Goodrich Co.).
123a. W. J. Kroenke, in 8th Akron Polymer Conference, Akron, Ohio, September 1977.
123b. W. J. Kroenke, *J. Appl. Polymer. Sci.*, **26**, 1167 (1981).
124. A. P. Wilson, U.S. Pat. 3,935,142 (1976) (to Standard Oil Co.).
125. H. P. Doerge and M. Wismer, U.S. Pat. 3,746,664 (1973); Ger. Pat. 2,053,731 (to PPG Ind.) (1971).
126. H. P. Doerge and M. Wismer, *J. Cell. Plast.* **8**, 311 (1972).
127. H. P. Doerge and M. Wismer, U.S. Pat. 3,639,307; Ger. Pat. 2,208,951 (to PPG Ind.) (1972).
128. H. P. Doerge, and M. Wismer, U.S. Pat. 3,647,724 (1972) (to PPG Ind.).
129. S. T. Kus and F. W. Koenig, U.S. Pat. 3,637,543 (1972) (to Standard Oil Co., Ind.).
130. K. Ashida, M. Ohtani, T. Yokoyama, and S. Ohkubo, *J. Cell. Plast.* **14**, 200 (1978).
131. K. C. Frisch, U.S. Pat. 3,725,319 (1973); Brit. Pat. 1,309,226 (1973); Ger. Pat. 2,121,821 (1971) (to Owens-Corning Corp.).
132. D. C. Wysocki, Ger. Pat. 2,422,647 (1975) (to Owens-Corning Corp.).
133. D. C. Wysocki, U.S. Pat. 3,933,694 (1976) (to Owens-Corning Corp.).
134. D. Hanton, U.S. Pat. 3,915,905 (1975) (to Certain-Teed Prods. Corp.).
135. R. Majchrowski and G. S. Wentworth, Can. Pat. 894,862 (1972) (to Canadian Pittsburgh, Ltd.).
136. Y. Ogawa, H. Hisada, H. Kimoto, N. Shiina, and T. Takenaka, Japan. Pat. 75:100147 (1975) (to Marubishi Oil Chem. Co., Ltd.) (CA 84:60511m).
137. T. Okina, Japan. Pat. 77:132098 (1977) (to M.T.P. Kasei Co., Ltd.) (CA 88:38536j).
138. E. D. Dickens, in Polymer Conference Series, University of Detroit, Detroit, Michigan, May 1975.
139. S. Kaufman and R. S. Dedier, "PVC Jacket Compound with Improved Flame Retardancy

and Superior Physical Properties" in Proceedings, 23rd International Wire & Cable Symposium, Cherry Hill, New Jersey p. 281 (published by NTIS, Springfield, Va.) (1974).
140. J. S. Heckles, U.S. Pat. 4,085,081 (1978) (to Armstrong Cork Co.).
141. T. C. Mathis and J. D. Hinchen, *SPE Tech. Papers* **19**, 343 (1973); T. C. Mathis and A. W. Morgan, Ger. Pat. 2,363,311 (1974) (to Monsanto Co.).
142. T. C. Mathis, G. W. Mappes, and C. M. Calvert, *SPE Tech. Papers* **23**, 190 (1977).
143. K. Ashida, M. Ohtani, O. Masaaki, and S. Ohkubo, Japan. Pat. 76:106198 (1976) (to Mitsubishi Chem. Ind. Co., Ltd.) (CA 86:17915a).
144. K. Iwasaki, Japan. Pat. 75:78699 (1975) (to Kokoku Chem. Ind. Co., Ltd.).
145. H. L. Vandersall, *J. Fire Flammability* **2**, 97 (1971).
146. F. J. O'Neill, U.S. Pat. 4,005,241 (1977) (to Monsanto Co.).
147. C. Ancolik, L. R. Perkins, C. W. Stewart, J. R. Galloway, and W. Hellyer, in Annual Convention, American Hospital Association, Atlanta, Georgia, 1977).
148. M. Shitara, H. Kokame, Y. Oguri, T. Muroi, and T. Yoshida, Japan. Pat. 73:51086 (1973) (to Hitachi, Ltd.).
149. J. M. Makhlouf and E. E. Parker, Ger. Pat. 2,714,039; U.S. Pat. 4,041,003 (1977) (to P.P.G. Ind., Inc.).
150. H. J. Fabris, Polymer Conference Series, University of Detroit, Detroit, Michigan, May 1974.
151. A. Tooya and T. Minegishi, Japan. Pat. 75:128795 (1975) (to Nippon Spring Mfg. Co., Ltd.) (CA 84:136624y).
152. T. Saito, T. Kaneda, and N. Kajitani, Japan. Pat. 77:42598 (1977) (to Nippon Soda Co., Ltd.).
153. K. Iwasaki, Can. Pat. 1,030,299 (1978) (to Kokoku Chem. Ind., Ltd.).
154. A. J. Papa, *Ind. Eng. Chem. Prod. Res. Develop.* **11**, 379 (1972).
155. W. J. Gilwee, R. W. Rosser, and J. A. Parker, "High Temperature Resistant Composites from Bismaleimide Resin" in Proceedings of the 18th National SAMPE Symposium, Los Angeles, California (published by Society for Advancement of Materials Process Engineering, Azusa, Cal.) (1973).
156. D. A. Kourtides and J. A. Parker, Pub. U.S. Pat. Appl. 778,195 (1977) (to United States National Aeronautics and Space Administration).
157. D. A. Kourtides, W. J. Gilwee, and J. A. Parker, in Regional Technical Conference, Connecticut Section, SPE, p. 113 (1977).
158. D. A. Kourtides and J. A. Parker, *Plast. Des. Process.* **18**, 53 (1978).
159. M. M. O'Mara, W. Ward, D. P. Knechtges, and R. J. Meyer, in *Flame Retardancy of Polymeric Materials* (W. C. Kuryla and A. J. Papa, eds.), Vol. 1, p. 217, Marcel Dekker, New York (1973).
160. M. M. O'Mara, *J. Polym. Sci.* **9**, 1387 (1971).
161. W. Pilz and A. Gatzke, *Res Discl.* **154**, 50 (1977).
162. E. J. Quinn, D. H. Ahlstrom, and S. A. Liebman, *Polym. Prepr.* **14**(2), 1022 (1973).
163. M. M. O'Mara, *Polym. Prepr.* **14**(2), 1028 (1973).
164. R. E. Singler, N. S. Schneider, and G. L. Hagnauer, *Polym. Eng. Sci.* **15**, 321 (1975).
165. D. F. Lawson and T. C. Cheng, *Fire Res.* **1**, 223 (1978).
166. R. E. Proodian, *SPE Tech. Papers* **19**, 714 (1973).
167. J. E. Thompson and K. A. Reynard, *J. Appl. Polym. Sci.* **21**, 2575 (1977).
168. K. A. Reynard, R. W. Sicka, J. E. Thompson, and S. H. Rose, Horizons, Inc., Cleveland, Ohio, Naval Ship Engineering Center Contract, N00024-73-5474, March 1975.
169. A. E. Oberster and D. F. Lawson, U.S. Pat. 4,064,095 (1977) (to The Firestone Tire and Rubber Co.).
170. R. L. Dieck and E. J. Quinn, U.S. Pat. 4,026,838 (1977) (to Armstrong Cork Co.).
171. E. J. Quinn and R. L. Dieck, *J. Fire Flammability* **7**, 5 (1976).
172. E. J. Quinn and R. L. Dieck, *J. Cell. Plast.* **13**, 96 (1977).
173. K. Sebata, J. H. Magill, and Y. C. Alarie, *J. Fire Flammability* **9**, 50 (1978).

174. C. L. Kehr, R. M. Murch, and N. S. Marans, U.S. Pat. 3,897,372 (1975) (to W. R. Grace Co.).
175. R. V. Russo, Ger. Pat. 2,539,982 (1976) (to G.A.F. Corp.).
176. A. H. Syrop, *Adv. Urethane Sci. Technol.* **4**, 176 (1975).
177. A. H. Syrop, *Int. Prog. Urethanes*, 167 (1977).
178. J. S. Sedlak and R. J. Raynor, in *Fire Retardants, Proceedings of International Symposium on Flammability and Fire Retardants, 1976*, p. 228, Technomic Publishing Co., Westport, Connecticut (1977).
179. T. Yukuta, T. Ohashi, A. Takashi, and K. Arai, Ger. Pat. 2,621,582 (1976) (to Bridgestone Tire Co., Ltd.).
180. T. Yukuta, T. Ohasahi, M. Kojima, and M. Saito, Japan. Pat. 77:104598 (1977) (to Bridgestone Tire Co., Ltd.) (CA 88:75072a).
181. D. F. Lawson and J. W. Kang, U.S. Pat. 4,005,248 (1977) (to The Firestone Tire and Rubber Co.).
182. D. F. Lawson, in 8th Central Regional ACS Meeting, Akron, Ohio, May 1976.
183. D. A. Alia, U.S. Pat. 3,971,882 (1976) (to Okonite Co.).
184. D. Florence, Ger. Pat. 2,360,950 (1974) (to Armstrong Cork Co.).
185. T. Murata, Japan. Pat. 78:00251 (1978) (to Matsushita Elect. Ind. Co., Ltd.) (CA 88: 171124q).
186. S. Kawawata and N. Takahata, *Japan. Plast. Age* **43** (September–October, 1975) (CA 84: 18269p).
187. Y. Ogawa, H. Hisada, K. Kimoto, N. Shiina, and H. Okutani, Japan. Pat. 75:100146 (to Marubishi Oil Chem. Co., Ltd.) (CA 83:207134f).
188. W. J. Kroenke, U.S. Pat. 3,922,248 (1975) (to B. F. Goodrich Co.).
189. W. J. Kroenke, U.S. Pat. 3,878,167; 3,914,201 (1975) (to B. F. Goodrich Co.).
190. F. W. Moore and G. A. Tsigdinos, Ger. Pat. 2,736,513 (1978) (to Amax, Inc.).
191. G. A. Tsigdinos, F. W. Moore, V. J. Kuck, and E. Scalco, Ger. Pat. 2,736,800 (1978) (to Amax, Inc.).
192. F. W. Moore and G. A. Tsigdinos, *J. Less-Common Met.* **54**, 297 (1977); D. A. Church and F. W. Moore, *Plast. Eng.* **31**(12), 36 (1975).
193. W. J. Kroenke, U.S. Pat. 4,053,455 (1977) (to B. F. Goodrich Co.).
194. M. I. Jacobs and C. F. Paddock, Ger. Pat. 2,654,666; M. I. Jacobs, Belg. Pat. 85542 (1977) (to Uniroyal, Inc.).
195. E. L. White, W. E. Robertson, and J. M. Schwarcz, U.S. Pat. 3,996,142 (1976) (to NL Ind., Inc.).
196. Belg. Pat. 841,616 (1976) (to NL Ind., Inc.).
197. J. M.-G. Huang, R. J. Capwell, and M. A. DeSesa, Ger. Pat. 2,743,624 (1977) (to NL Ind., Inc.).
198. Neth. Pat. 76 05250 (1977) (to NL Ind., Inc.).
199. E. L. White, J. M. Schwarcz, and B. Engelhardt, Ger. Pat. 2,460,194 (1975) (to NL Ind., Inc.).
200. E. L. White, W. E. Robertson, and J. M. Schwarcz, Ger. Pat. 2,620,092 (1977) (to NL Ind., Inc.).
201. Y. Konoshita, A. Yamamoto, A. Kitagawa, and M. Shindo, Japan. Pat. 74:77967 (1974) (to Nissan Corp.) (CA 82:4429u).
202. P. Kraft and L. Smalheiser, Fr. Pat. 2,192,562; G.B. Brit. Pat. 1,382,625 (1975) (to Stauffer Chem. Co.).
203. A. W. McRowe, U.S. Pat. 3,819,577 (1974) (to B. F. Goodrich Co.).
204. A. W. McRowe, Ger. Pat. 2,700,656; Fr. Pat. 2,337,746 (1977) (to B. F. Goodrich Co.).
205. Tech. Data Sheet No. 287 (M.&T. Chem., Inc.).
206. J. J. Kostrowski and W. A. Larkin, Fr. Pat. 2,071,855 (1971); Ger. Pat. 2,054,304 (1972) (to M.&T. Chem., Inc.).
207. Data Sheet F-4, NL Ind., Inc.

208. T. Bito and T. Saigo, Japan. Pat. 76:23548, (1976) (to Teijin, Ltd.) (CA 84:181196w).
209. W. J. Kroenke, U.S. Pat. 3,883,480; U.S. Pat. 3,883,482 (1975) (to B. F. Goodrich Co.).
210. H. Stone and A. B. Robertson, U.S. Pat. 3,812,066 (1974); Ger. Pat. 2,230,159 (1972) (to Allied Chem. Co.).
211. H. Yamashita, H. Kaminoyama, N. Sawada, and K. Tsurumaki, Japan. Pat. 76:17949 (1976) (to Fujikura Cable Works, Ltd.) (CA 84:151633u).
212. D. F. Lawson and E. L. Kay, U.S. Pat. 4,035,324 (1977) (to The Firestone Tire and Rubber Co.).
213. H. Kitayama, *Raba Daijesuto* **28**(6), 11 (1976) (CA 86:59901y).
214. L. C. Mitchell, U.S. Pat. 3,846,372 (1974) (to Ethyl Corp.).
215. D. F. Lawson, U.S. Pat. 4,096,115 (1978) (to The Firestone Tire and Rubber Co.).
216. S. R. Sandler, U.S. Pat. 4,011,194 (1977) (to Pennwalt Corp.).
217. S. R. Sandler, U.S. Pat. 4,049,612 (to Pennwalt Corp.).
218. D. F. Lawson, *Res. Discl.* **161**, 4 (1977).
219. Y. Yamamotu and N. Watani, Japan. Pat. 78:56248 (1978) (to Showa Kako K. K.) (CA 89: 111522d).
220. J. E. Hartitz, Belg. Pat. 859763 (1978) (to B. F. Goodrich Co.).
221. E. D. Dickens, Fr. Pat. 2,346,401; U.S. Pat. 3,983,086 (1977) (to B. F. Goodrich Co.).
222. E. D. Dickens, U.S. Pat. 3,962,177; U.S. Pat. 3,965,068; U.S. Pat. 3,968,081; U.S. Pat. 3,968,064; U.S. Pat. 3,970,638; U.S. Pat. 3,975,356; U.S. Pat. 3,975,359 (1976) (to B. F. Goodrich Co.).
223. W. J. Kroenke, U.S. Pat. 4,053,452; U.S. Pat. 4,053,454; U.S. Pat. 4,055,537; U.S. Pat. 4,055,538 (1977) (to B. F. Goodrich Co.).
224. W. J. Kroenke and A. W. McRowe, U.S. Pat. 4,053,451 (1977) (to B. F. Goodrich Co.).
225. A. W. McRowe and W. J. Kroenke, U.S. Pat. 4,053,453 (1977) (to B. F. Goodrich Co.).
226. J. E. Hartitz and R. A. Yount, *Polym. Eng. Sci.* **18**, 549 (1978).
227. A. W. Morgan and T. C. Mathis, Ger. Pat. 2,657,999 (1977) (to Monsanto Co.).
228. D. A. Alia, U.S. Pat. 3,821,139 (1974) (to Okonite Co.).
229. A. W. Morgan and T. C. Mathis, U.S. Pat. 4,083,789 (1978) (to Monsanto Co.).
230. L. Parts, R. D. Myers, C. A. Thompson, and N. F. May, U.S. NTIS, AD Rep., AD-A049923 (1978).
231. P. V. Bonsignore and T. L. Levendusky, *J. Fire Flammability* **8**, 95 (1977).
232. S. J. Chlystek, U.S. Pat. 4,053,439 (1977) (to Armstrong Cork Co.).
233. H. P. Doerge and M. Wismer, U.S. Pat. 3,637,542 (1972) (to PPG Ind., Inc.).
234. W. J. Kauffman and E. J. Quinn, Ger. Pat. 2,718,063 (1977) (to Armstrong Cork Co.).
235. K. Ashida, M. Ohtani, and S. Ohkubo, Ger. Pat. 2,609,181; Ger. Pat. 2,610,640 (1976); Japan. Pat. 77:109598 (1977) (to Mitsubishi Chem. Ind. Co., Ltd.) (CA 88:171076a).
236. K. Ashida, M. Ohtani, T. Yokoyama, and S. Ohkubo, Ger. Pat. 2,740,093 (1978) (to Mitsubishi Chem. Ind. Co., Ltd.).
237. Neth. Pat. 76:09427; 76:09429 (1978) (to Mitsubishi Chem. Ind. Co., Ltd.) (CA 88:171167f, 25388h).
238. K. Ashida, M. Ohtani, T. Yokoyama, and S. Ohkubo, Ger. Pat. 2,542,499 (1976) (to Mitsubishi Chem. Ind. Co., Ltd.).
239. R. Weidermann, R. Merten, W. Dietrich, and W. Schmitz, Ger. Pat. 2,637,170 (1978) (to Bayer A.-G.).
240. R. A. Kolakowski, H. E. Reymore, and A. A. R. Sayigh, Ger. Pat. 2,656,600 (1977) (to Upjohn Co.).
241. P. Merriman, C. V. Keen, and R. W. H. Bell, Ger. Pat. 2,213,127 (1972) (to Dunlop Ltd.).
242. Y. Hiresaki and M. Matsumura, Japan. Pat. 75:74696 (1975) (to Bridgestone Tire Co., Ltd.).
243. T. Ohashi, T. Okuyama, K. Arai, Y. Taniguchi, M. Saito, and R. Sakata, Japan. Pat. 78:16097 (1978); Ger. Pat. 2,734,400 (1978) (to Bridgestone Tire Co., Ltd.) (CA 89:111889d).
244. R. C. Stauffer, E. R. Larsen, R. V. Petrella, A. Manca, and D. P. Miller, *J. Fire Retard. Chem.* **3**, 34 (1976).

245. O. K. Goins and Y. C. Chae, in 27th Annual Technical Conference, SPI, Paper 19-A (1972).
246. B. J. Bircher, Ger. Pat. 2,607,041 (1976) (to ICI, Ltd.).
247. D. E. Stuetz and C. L. Smart, U.S. Pat. 3,508,872 (1970) (to Celanese Corp.).
248. E. Dorfman, R. R. Hindersinn, and W. T. Schwartz, U.S. Pat. 3,983,185; Ger. Pat. 2,531,816 (1976) (to Hooker Chem. and Plast. Co.).
249. E. Dorfman, W. T. Schwartz, and R. R. Hindersinn, U.S. Pat. 4,013,815 (1977) (to Hooker Chem. & Plast. Co.).
250. Y. Uemura, Japan. Pat. 74:07572 (1974) (CA 81:106734e).
251. G. A. Skinner, L. E. Parker, and P. J. Marshall, *Fire Mater.* **1**, 154 (1976).
252. H. P. Doerge and M. Wismer, U.S. Pat. 3,758,638 (1973) (to PPG Ind.).
253. A. J. Spak and A. J. Yu, U.S. Pat. 4,088,710 (1978) (to Stauffer Chem. Co.).
254. L. E. Shea, U.S. Pat. 4,053,447 (1977).
255. T. F. Jonas, *National SAMPE Symposium and Exhibition* **21**, 1015 (1976).
256. D. F. Lawson, *Res. Discl.* **160**, 10 (1977).
257. C. W. Stewart, R. L. Dawson, and P. R. Johnson, *Rubber Chem. Technol.* **48**, 132 (1975).
258. D. F. Lawson and E. L. Kay, U.S. Pat. 4,059,560 (1977) (to The Firestone Tire and Rubber Co.).
259. S. Nogami and K. Waki, Japan. Pat. 73:43076 (1973) (to Asahi Chem. Ind. Co., Ltd.) (CA 79:67311b).
260. F. W. Moore, C. J. Hallada, and H. F. Barry, U.S. Pat. 3,956,231 (1976) (to Amax, Inc.).
261. D. F. Lawson, *Res. Discl.* **160**, 7 (1977).
262. L. P. Parts and J. T. Miller, U.S. Pat. 3,825,520 (1974) (to Monsanto Research Corp.).
263. L. P. Parts and J. T. Miller, U.S. Pat. 3,766,157 (1974) (to Monsanto Research Corp.).
264. L. P. Parts and J. T. Miller, U.S. Pat. 3,793,404 (1974) (to Monsanto Research Corp.).
265. F. C. Weissert, U.S. Pat. 2,880,183 (1959) (to The Firestone Tire and Rubber Co.).
266. A. R. Ingram, U.S. Pat. 4,035,315 (1977) (to Arco Polymers, Inc.).
267. Japan. Pat. 77:124045 (1977) (to Monsanto Co.) (CA 89:111303h).
268. Y. Toyosaki, T. Natsuume, and E. Iida, Japan. Pat. 74:07349 (1974) (CA 81:14313p).
269. H. Asai and T. Natsuume, Japan. Pat. 74:44053 (1974) (CA 81:137075p).
270. C. Savides, A. Granzow, and J. F. Cannelongo, in Technical Papers, Regional Technical Conference, SPE, 18 (1975) (CA 83:165007j).
271. K. Yagi and M. Yamamoto, U.S. Pat. 4,001,174 (1977) (to Yamazaki Corp.), 1977.
272. T. Tsumori, Y. Ikegami, K. Tsukagoe, K. Fukatsu, and Y. Okabe, Japan. Pat. 75:24531 (1975) (to Kohjin Co., Ltd.) (CA 83:81065P).
273. T. Tsumori and Y. Okabe, Japan. Pat. 75:29682 (1975) (to Kohjin Co., Ltd) (CA 82:80757d).
274. F. J. Martin and K. R. Price, *J. Appl. Polym. Sci.* **12**, 143 (1968).
275. P. V. Bonsignore, *SPE Tech. Papers* **22**, 472 (1976).
276. P. V. Bonsignore and J. H. Manhart, *Kunststoffe* **67**(1), 34 (1977).
277. P. V. Bonsignore, *Plast. Eng.* **32**(10), 41 (1976).
278. T. Yagi, Japan. Pat. 74:21043; 74:21042 (1974) (to Kishimoto, Sangyo Co., Ltd.) (CA 82:75181m, 99543g).
279. R. L. Dieck and E. J. Quinn, U.S. Pat. 4,083,820 (1978) (to Armstrong Cork Co.).
280. O. Yoshikatsu, H. Haruhiko, H. Kimoto, and T. Takenaka, Japan. Pat. 75:89605 (1975) (to Marubishi Oil Chem. Co., Ltd.) (CA 84:6784q).
281. J. J. Brenden, U.S. Forest Service, Res. Pap. FPL 249 (1975).
282. Y. Ogawa, H. Hisado, N. Shiina, and H. Okutani, Japan. Pat. 75:89511 (1975) (to Marubishi Oil Chem. Co., Ltd.) (CA 84:6756g).
283. T. Hanada and M. Saito, Japan. Pat. 73:05895 (1973) (to Adachi Veneer Co., Ltd.) (CA 80:16693r).
284. A. Hamoudi, M. Bert, and A. Guyot, *Fire Res.* **1**, 101 (1977).
285. Fr. Pat. 2,172,797 (1973) (to Gamlen Naintre, S.A.).
286. K. Shimizu and K. Suzuki, U.S. Pat. 3,663,355 (1972).

287. L. Benisek and W. A. Phillips, *J. Fire Flammability* **8**, 458 (1977).
288. L. Benisek and W. A. Phillips, *J. Fire Flammability* **8**, 247 (1977).
289. G. Coppa-Zuccari, *Gummi Asbest Kunstst.* **29**(4), 208 (1976).
290. L. L. Fewell, E. L. Trabold and H. H. Spieth, *J. Fire Flammability* **9**, 377 (1978).
291. E. L. Yuan, U.S. Pat. 3,811,997 (1974) (to E. I. duPont de Nemours and Co.).
292. R. L. Dieck and E. J. Quinn, U.S. Pat. 4,026,839 (1977) (to Armstrong Cork Co.).
293. E. J. Quinn and R. L. Dieck, *J. Fire Flammability* **7**, 358 (1976).
294. R. L. Dieck and E. J. Quinn, *J. Fire Flammability* **8**, 412 (1977).
295. R. L. Dieck and E. J. Quinn, U.S. Pat. 4,061,606 (1977) (to Armstrong Cork Co.).
296. R. L. Dieck and E. J. Quinn, U.S. Pat. 4,053,456 (1977) (to Armstrong Cork Co.).

3

*Experimental Evaluation of Flammability Parameters of Polymeric Materials**

Archibald Tewarson

Abstract

An evaluation of the fire behavior of polymers and liquids over a wide range of experimental conditions is made using a laboratory scale flammability apparatus developed by the author. Results are presented for the following fuel parameters: (1) minimum heat flux (surface temperature), energy, and critical mass loss rate required for the piloted ignition of fuel vapor–air mixture and kinetic parameter for fuel vapors; (2) "effective" heat of gasification of the fuel; (3) flame-convective and flame-radiative heat flux to the fuel surface; (4) mass generation rates of CO, CO_2, gaseous hydrocarbons, and "pyrolyzate",† expressed as fractional theoretical stoichiometric yields (or fractions of carbon in the fuel converted to the products); (5)

*This report contains information derived from a research project sponsored by the Products Research Committee. However, any conclusions drawn from the research project in this report are those of the author and not of the PRC.

†Pyrolyzate' is defined as a mixture of carbon-containing products other than CO, CO_2, and gaseous hydrocarbons.

Archibald Tewarson • Factory Mutual Research Corporation, Factory Mutual System, Norwood, Massachusetts 02062.

chemical formula of the fuels based on measured elemental compositions; (6) heat release rates (actual, convective, and radiative) expressed as combustion efficiency of the fuel vapors and convective and radiative fractions of the theoretical stoichiometric heat release rate for the complete combustion of the fuel vapors; (7) net heat of complete combustion and actual heat of combustion of the fuels; and (8) the ratio of optical density per unit path length to mass concentration of the fuel vapors defined as "modified mass absorbency index."

The apparatus and concepts used for obtaining the fuel parameters are also described.

The parameters are obtained for fuels about 0.008 m^2 in area exposed to various values of external heat flux and mass fraction of environmental oxygen.

The results are reported for the following fuels*:

1. *Aliphatic-Type Fuels*
   a. Liquids: methanol, acetone, heptane.
   b. Granular: cellulose, polyoxymethylene, polyethylene, polypropylene, polymethylmethacrylate, nylon.
   c. Foams: polyethylene.
2. *Aromatic-Type Fuels*
   a. Liquids: aniline, benzene, styrene.
   b. Granular: polystyrene, styrene–butadiene.
   c. Foams: polystyrene†, polyurethanes†, polyisocyanurates†, phenolic.
3. *Chlorinated-Type Fuels.* Chlorinated polyethylenes, polyvinyl chloride.

Results for composite fuels are also reported.

This report is based on work which was started in July 1975 and completed in November 1978 under Products Research Committee Grant No. RP-75-1-33A.

Objectives

The objectives of the author's research on the flammability of fuels are (1) to obtain data for the fuel parameters over a wide range of experimental conditions so that the data are useful as input parameters for fire modeling and for making engineering decisions as to the safe applications of the polymers for various end uses; (2) to establish a simplified laboratory scale test method and technique for the flammability evaluation of polymers so that manufacturers, consumers, and regulatory agencies can use them for deter-

*Some of the fuels contain fire retardants. Some fires are not fully ventilated.

† Samples are standard reference samples of the Products Research Committee and are identified as GM-21, 23, etc.

mining the fire hazard, for improving the fire safety of the fuels, and for regulations and codes; (3) to define the fire hazard for the end-use applications of the polymers and to establish a methodology for a fire hazard index (or indices); (4) to use established scientific principles to understand the fire behavior of fuels so that further improvements can be made in fire modeling and for the hazard evaluation of the fuels in real fires.

In the author's continuing research three similar flammability apparatuses are involved, employing samples areas of about 0.008, 0.03, and 0.09 m^2. The measurements made in each apparatus are identical. In addition, available data for larger scale fires are also used.

In the following presentation, results are presented for the fuel parameters using samples with areas of about 0.008 m^2 exposed to various values of external heat flux and mass fraction of the environmental oxygen.

1. Introduction

A laboratory-scale flammability apparatus has been developed by the author to evaluate the fire behavior of polymers and liquids (defined as fuels) over a wide range of experimental conditions. The fire behavior of fuels is expressed in terms of ignition, mass loss rate of the fuels in pyrolysis and combustion, mass generation rates of pyrolysis and combustion products, actual, convective, and radiative heat release rates, optical transmission (or visibility) through the products, generation of toxic compounds, and extinction of the fire.

The flammability apparatus (Fig. 1) uses a sample about 0.008 m^2 in area, about 0.02–0.05 m thick. Experiments can be performed under natural convection or under forced convection (a quartz tube surrounds the sample, and air with variable flow rates and oxygen concentration is introduced at the bottom of the apparatus.) The sample is exposed to different magnitudes of external heat flux ranging from 0 to 70 kW/m^2.

All the products generated in the pyrolysis and combustion of the fuel are captured in a sampling duct, where the dilution of the products with fresh air is variable (maximum dilution is about 10 times by volume).

In the experiments the variables include (1) type of the fuel, (2) air flow rate entering the apparatus, (3) concentration of oxygen in the air entering the apparatus, (4) external heat flux applied to the sample, (5) volumetric ratio of the products to fresh air captured in the sampling duct.

The measurements made in the experiments include (1) time to piloted ignition or autoignition; (2) mass loss in the pyrolysis and combustion; (3) sample surface temperature; (4) temperature of the mixture of products and air captured in the sampling duct; (5) concentrations of CO, CO_2, O_2, and total

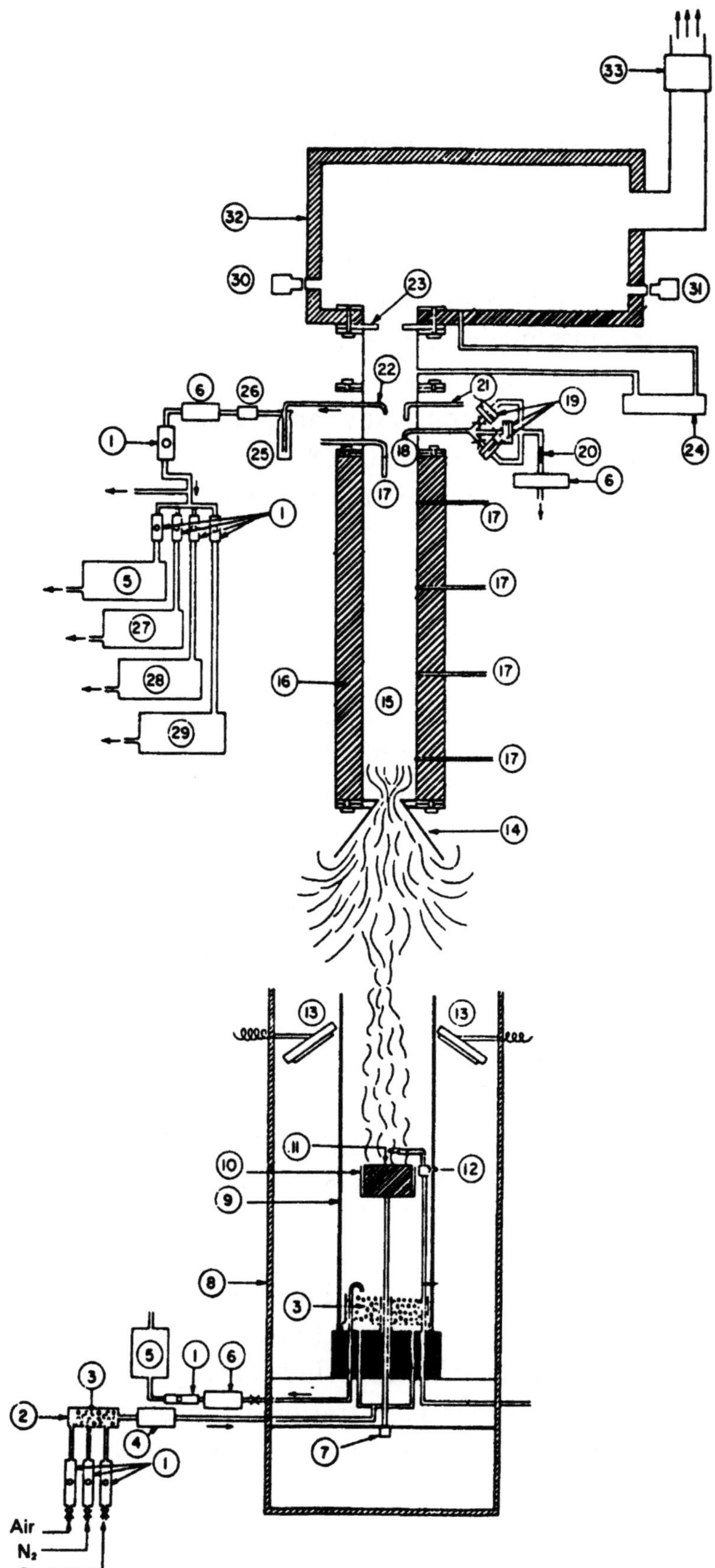

FIGURE 1. Laboratory scale flammability apparatus.

gaseous hydrocarbons (as CH_4), other gaseous, liquid, and solid carbon-containing or non-carbon containing products; (6) total volumetric and mass flow rate of the mixture of products and air captured in the sampling duct; (7) optical transmission through the mixture of products and air captured in the sampling duct; and (8) concentrations of gaseous agents such as N_2, CO_2, halons or the volumetric or mass flow rate of liquid agents such as water required for the extinction of the fire.

The apparatus has also been designed so that animals could be exposed to products generated in the pyrolysis or combustion of the fuels.

The elemental composition and heat of combustion of the fuels are measured. The elemental composition of the fuel is used to calculate the theoretical stoichiometric mass of each product produced per unit mass of the fuel converted completely to the product, k_i. The elemental composition is also used to calculate the theoretical stoichiometric mass of oxygen consumed per unit mass of fuel (k_{O_2}).

The ease of ignition of the fuel is expressed in terms of minimum heat flux ($\dot{q}''_0$), energy (E_{eff}), and critical mass loss rate ($\dot{m}''_c$) required for the ignition of fuel vapor–air mixture, where ignition is either piloted ignition or autoignition. The mass loss rate in the pyrolysis of the fuel is expressed in terms of "effective" heat of gasification of the fuel (L_g) and "effective" surface reradiation loss ($\dot{q}''_0$). The mass loss rate in the combustion of the fuel is expressed in terms of the ratio of flame heat flux to the surface to effective heat of gasification. The flame heat flux to the surface can be predominantly convective or predominantly radiative depending on the experimental conditions used. However, for laboratory scale data to be useful for predicting the behavior of fuels in large-scale fires, the ratio of flame radiative heat flux to the surface to effective heat of gasification, $\dot{q}''_{fr}/L_g$, is one of the most useful parameters.

The mass generation rates of the products, $\dot{M}''_i$, are expressed in terms of the fractional theoretical stoichiometric yields of the products, or the fraction of the carbon or other elements in the fuel converted to individual products ($f_i = \dot{M}''_i/k_i \cdot \dot{m}''_b$, where k_i is the theoretical stoichiometric mass of the product produced per unit mass of the fuel converted completely to the product, and $\dot{m}''_b$ is the mass loss rate in the combustion of the fuel).

The convective, radiative, and actual heat release rates, $\dot{Q}''_i$, are expressed as fractions of the theoretical stoichiometric heat release rate for complete combustion of the fuel, X_i. The theoretical stoichiometric heat release rate for complete combustion of the fuel is given by $\dot{m}''_b \cdot H_c$, where $\dot{m}''_b$ is mass loss rate in the combustion of the fuel and H_c is net heat of complete combustion of the fuel.

The optical transmission (or visibility) through the mixture of products and air is expressed in terms of the "modified mass absorbency index," ξ_{pyr}, given as the ratio of the optical density per unit path length to the mass concentration of the fuel vapors. Optical density per unit path length is given by $(1/l)\ \ln\ (I_0/I)$, where l is optical path length and I_0 and I are optical transmission through air and through the mixture of products and air, respectively.

The author has not studied the extinction of the fire and generation of toxic compounds (including animal experiments) so far, and thus the concepts have not been defined here.

In this report, data for the fuel parameters have been presented for widely varying conditions of external heat flux ($\dot{q}''_e$) and mass fraction of oxygen (m_{O_2}) under forced convection; the air flow rate entering the apparatus and total flow rate of mixture of products and air leaving the apparatus have been maintained constant. Data for only the quasi-steady-state conditions for pyrolysis and combustion of the fuels are presented. Data are presented in five sections dealing with individual fuel parameters. Each section also contains the concept for individual fuel parameters.

2. *Experimental Procedures*

The laboratory scale flammability apparatus is shown in Fig. 1; each component of the apparatus is described separately.

2.1. *Sample and Sample Container*

The sample used in the study is about 0.008 m^2 in area, and about 0.02–0.05 m in thickness and is placed at location No. 11 and Fig. 1. The sample container (No. 10) is made from two sheets of heavy duty aluminum foil (Reynolds Wrap).

2.2. *Total Flow Rate of Mixture of Air–O_2–N_2 and Mass Fraction of Oxygen*

The sample is enclosed inside an air-tight quartz tube (No. 9) about 0.17 m in diameter and about 0.61 m long. The total flow rate of the mixture of air–O_2–N_2 entering the quartz tube is about 1.4×10^{-3} m^3/sec and is maintained constant in all the experiments. The flow rate is monitored by an electronic flow meter (No. 4, Model 2012, Thermo System Inc., St. Paul, Minnesota).

The mass fraction of oxygen in the mixture of air–O_2–N_2 entering the quartz tube is monitored by a precalibrated oxygen analyzer (No. 5, Beckman, Model F-3). In the experiments the mass fraction of oxygen is varied from 0.0 to about 0.60, depending on the type of experiment being performed.

The mixture of air–O_2–N_2 is passed through two drying tubes (anhydrous $CaSO_4$, Hammond Co., Xenia, Ohio) before entering the apparatus.

2.3. External Heat Flux

Four high-density radiant heaters each containing six tungsten quartz lamps (No. 13, Model 5208, Research Inc., Minneapolis, Minnesota) are used coaxially for the application of external heat flux to the sample. The output of all four radiant heaters is controlled by a single controller (Model 646, Research, Inc., Minneapolis, Minnesota).

The radiant heat flux to the sample surface is calibrated by using a heat flux gauge (Model No. 16-10-GTW64, Medtherm Corporation, Huntsville, Alabama). The heat flux gauge is traversed horizontally and vertically within the range of the sample dimension (0.05 m high, 0.1-m diameter). Each radiant heater is adjusted by forward and backward or upward and downward movements and by varying the angle of inclination until the heat flux within the range of the sample dimension is reasonably uniform. For each controller setting, the heat flux gauge is placed at about a hundred locations within the range of the sample dimension, and the steady state output of the heat flux gauge is recorded for each location. All the output values are averaged and multiplied by the calibration constant of the gauge (supplied by the Medtherm Corporation). The deviation between various values is about ±5% of the average value. This procedure is repeated for different controller settings covering the range from 10 to 70 kW/m^2 at the sample surface. A calibration curve for the radiant heaters is then constructed where the average heat flux values at the sample surface are plotted as functions of the controller settings. In the experiments, the calibration curve then is used to determine the heat flux value being applied to the sample by noting the controller setting.

In all the experiments, the radiant heaters are stabilized by keeping them on for 300 sec prior to exposing the sample to the radiant heat flux.

2.4. Ignition

For piloted ignition, a premixed hydrocarbon–air burner is used (No. 12). The flame is about 0.01 m in length, and is located at the edge of the sample, about 0.01 m above the sample surface.

For the autoignition of the sample, experiments are very similar to the piloted ignition experiments, except for the absence of the flame.

2.5. Mass Loss Rate

The mass loss of the sample is measured by a water-cooled load cell (No. 7, Model UC3, Statham Instruments, Oxnard, California). Prior to each experiment, the output of the load cell is calibrated by standard weights (in the range of the sample weight).

2.6. *Total Mass Flow Rate of Mixture of Products and Air*

All the pyrolysis–combustion products are captured in the sampling duct (No. 15), together with fresh room air, diluting the products by about twenty times by volume. The sampling duct is designed such that the products and air are well mixed before the samples are taken for the analysis and heat losses are minimal.

The total mass flow rate of the mixture of hot products and air through the sampling duct is calculated from the following equation:

$$\dot{M}''_t = 9.16 \times [(p_a - \Delta p_d) \cdot \Delta p_m / T_d]^{1/2} \cdot (1/a) \tag{1}$$

where $\dot{M}''_t$ is the total mass flow rate of the mixture of hot products and air through the sampling duct per unit sample surface area (g/m^2 sec), p_a the measured ambient pressure (atm), Δp_d the measured pressure difference between inside and outside of the sampling duct (atm), Δp_m the measured pressure difference across the measuring orifice plate in the sampling duct, No. 23 (atm), T_d the measured gas temperature in the sampling duct, No. 17 (K), and a is the sample surface area (m^2).

2.7. *Convective Heat Release Rate*

The convective heat release rate is calculated from the following equation:

$$\dot{Q}''_{\text{conv}} = [\dot{M}''_t\, C_p(T)\,(T_d - T_a)]_{\text{exp}} - [\dot{M}''_t\, C_p\,(T)\,(T_d - T_a)]_{\text{blank}} \tag{2}$$

where, $\dot{Q}''_{\text{conv}}$ is the convective heat release rate per unit sample surface area (kW/m^2), C_p the specific heat of air of the mixture of products and air (kJ/g K), and $(T_d - T_a)$ is the measured gas temperature rise in the sampling duct above ambient temperature (K); exp denotes combustion experiments, blank identical blank experiment in the absence of combustion.

2.8. *Mass Generation Rates of Gaseous Pyrolysis–Combustion Products*

The concentrations of CO, CO_2, O_2, and total gaseous hydrocarbons (as CH_4) are measured by Beckman analyzers Nos. 5, 27, 28, 29 (Model 315A, IR analyzers for CO and CO_2; Model 755 paramagnetic analyzer for O_2; Model 400, flame-ionization analyzer for hydrocarbons).

The sample of the mixture of products and air is taken from location 22, and, before it is introduced into the analyzers, it is passed through two traps,

No. 25 in series, immersed in a cooling bath maintained at −30° C (Model RTE-8 Neslab Instruments, Portsmouth, New Hampshire). The traps contain glass beads on top of which glass wool is placed. After leaving the traps, the sample is introduced into a drying tube 26, (anhydrous $CaSO_4$, Hammond Co., Xenia, Ohio), and finally the sample is passed through a filter paper (type LS, 5 μm, Millipore Corporation, Bedford, Massachusetts). The sampling rate is about 2×10^{-4} m^2/sec.

The delay time (response time of the analyzer plus time for the sample to reach the analyzer) for each analyzer is measured by creating an appropriate gas environment near location 22, and by measuring the time taken by the analyzer for 90% of the full response; time zero is the time at which sampling is started.

For the calibration check of individual analyzers, gas environments of known concentrations of the calibration gas and pure dry N_2 are created separately near location 22.

Before starting the experiment, data for the background concentrations of gases are measured.

The mass generation rates of the pyrolysis–combustion products or mass depletion rate of O_2 are calculated from the following equation:

$$\dot{M}''_i = \Delta C_i \, \dot{M}''_T \, \delta_i \tag{3}$$

where, $\dot{M}''_i$ is the mass generation (or depletion) rate of the product per unit sample surface area (g/m^2 sec), ΔC_i the difference between initial and final concentration of the product (volume fraction), and δ_i the ratio of the density of the product to the density of air.

2.9. *Mass Generation Rate of The Pyrolyzate Fraction Collected on a Filter Paper*

The "pyrolyzate" is defined as a mixture of carbon-containing products other than CO, CO_2, and total gaseous hydrocarbons and is calculated by carbon balance.

The mass generation rate of the pyrolyzate fraction collected on a filter paper (Nos. 6, 18, 19, 20) is calculated from the following equation:

$$\dot{M}''_{pyr,f} = \frac{\Delta w}{\Delta t \dot{v}} \frac{\dot{V}_T}{a} \tag{4}$$

where $\dot{M}''_{pyr,f}$ is the mass generation rate of the pyrolyzate fraction collected on a filter paper per unit sample surface (g/m^2 sec), $\dot{v}$ the volumetric flow rate of the mixture of products and air through the filter paper (m^3/sec), Δw the mass of the pyrolyzate fraction collected on the filter paper (g), Δt the time during which pyrolyzate fraction is collected (sec), $\dot{V}_T$ the total volumetric flow rate of the mixture of products and air (m^3/sec), and a is the sample surface area (m^2).

2.10. Optical Transmission Through the Mixture of Pyrolysis-Combustion Products and Air

The optical transmission, for a fixed path length of 0.749 m, is measured in the flow direction of the mixture of hot products and air (Nos. 30 and 31). The light source used is a tungsten lamp (No. 1047, 26 V–2.70 A, operated at 20 V) and the light is received by a vacuum phototube (RCA 1P39). The collimating windows are fabricated from a planoconvex lens (Melles Griot 01LP XIII).

In order to reduce the pyrolyzate deposition on the windows of the optical system, air is introduced at the windows, at a rate of about 2×10^{-5} m^3/sec, (which is very small compared to the value of about 0.04 m^3/sec for the volumetric flow rate of the mixture of products and air flowing through the system).

The optical density per unit path length is calculated by the following equation:

$$D = \frac{1}{l} \ln (I_0/I) \tag{5}$$

where D is the optical density, l the optical path length (m), I_0 the optical transmission through ambient air (mV), and I is the optical transmission through the mixture of products and ambient air (mV).

3. Ignition

3.1. Concept

For ignition, the most commonly used relationship is the one developed for the surface temperature from the solution of linear flow of heat for a semiinfinite solid.[1]

$$T_s - T_a = \alpha \dot{q}''_e \left(\frac{4t_{ig}}{\pi k \rho C_p} \right)^{1/2} \tag{6}$$

where T_s is the surface temperature (K), T_a the ambient temperature (K), α the surface absorptance, $\dot{q}''_e$ the externally applied heat flux (kW/m^2), k the thermal conductivity (kW/m K), ρ the density (g/m^3), C_p the specific heat (kJ/g K), and t_{ig} is the ignition time (sec).

Equation (6) can be rearranged to give the following relationship:

$$t_{ig} = \frac{[(T_s - T_a)^2/2\alpha]\, \pi k \rho C_p}{(\dot{q}''_e)^2} \tag{7}$$

The experimentally measured t_{ig} values, for semiinfinite polymers, thus should show a linear relationship with values of $1/(\dot{q}''_e)^2$, the slope being equal to the numerator of Eq. (7), which can also be calculated from the known values of T_s, T_a, α, k, ρ, and C_p for the sample. In our studies, however, t_{ig} values measured for various polymers do not satisfy Eq. (7), probably for the following reasons: (1) $k\rho C_p$ and α are not constants, (2) heat loss from the surface due to reradiation is not negligible and, (3) phase changes as well as surface and condensed phase processes cannot be neglected. However, our data show that $t_{ig} \propto 1/\dot{q}''_e$, which can be expressed by the following equation:

$$\frac{1}{t_{ig}} = \frac{\dot{q}''_e - \dot{q}''_0}{E_{eff}} \tag{8}$$

where, $\dot{q}''_0$ is defined as the "effective" surface reradiation loss (kW/m^2) and E_{eff} the "effective" energy for ignition (kJ/m^2). E_{eff} is expected to contain the important properties for the ignition of the polymers including the apparatus-dependent parameters.

From $\dot{q}''_0$, the surface temperature can be estimated as follows:

$$T_s \cong \left(\frac{\dot{q}''_0}{\epsilon\sigma}\right)^{1/4} \tag{9}$$

where, ϵ is the surface emissivity (usually assumed to be unity) and σ is the Stefan–Boltzmann constant ($kW/m^2\ K^4$).

According to Eq. (8) for a fixed $\dot{q}''_e$ value, t_{ig} will be short if $\dot{q}''_0$ and E_{eff} values are small. If surface flame spread is treated as a series of ignitions of the fuel vapor–air mixtures created near the surface, then the smaller the values of $\dot{q}''_0$ and E_{eff}, the faster will be the surface flame spread. Thus, the advantages of the modification of the polymers such as surface coatings and chemical treatments (incorporation of the fire retardants, copolymerization, etc.) can be established by examining the increase in the magnitudes of $\dot{q}''_0$ and E_{eff}. The value of $\dot{q}''_0$ is insensitive to experimental conditions; however, E_{eff} is sensitive to variations in the experimental conditions. Thus, the influence of the variations in the experimental conditions on ignition can also be established by examining the variations in the E_{eff} values.

For the ignition of the polymer vapor–air mixture, a certain vapor production rate is required, which can also be measured in the ignition experiments. Rasbash,[2,3] defines the vapor production rate, for piloted ignition, in terms of a critical mass loss rate $\dot{m}''_c$. If it is assumed that the ignition process is a convective-heat-transfer-dominated process, then the relationship for a flame just at the critical condition for ignition is given by[2,3]

$$H_{fc} = \frac{m_{O_2} H_c/k_{O_2} - C_p(T_s - T_a)}{\exp(\dot{m}''_c C_p/h) - 1} \tag{10}$$

where H_{fc} is the heat transfer to the fuel by convection from the flame per unit mass of the fuel transported from the surface to the flame (kJ/g), m_{O_2} the mass fraction of environmental oxygen, H_c the net heat of complete combustion (kJ/g), k_{O_2} the theoretical stoichiometric mass of oxygen consumed per unit mass of the fuel, C_p the specific heat of air (kJ/g K), T_s the surface temperature (K), T_a the ambient temperature (K), $\dot{m}_c$ the critical mass loss rate per unit sample surface area (g/m^2 sec), and h is the convective heat transfer coefficient (kW/m^2 K).

The maximum fraction of heat of combustion which the flame reactions may lose to the sample surface by convection, without flame extinction, is defined by Rasbash[2,3] to be a kinetic parameter. The kinetic parameter, ϕ, is given by the following equation:

$$\phi = \frac{H_{fc}}{H_c} \tag{11}$$

ϕ is expected to be higher for fast-burning polymer vapors, and lower for slower-burning polymer vapors (polymers containing Cl, S, N, etc.).

3.2. *Measurements and Calculations*

The time to ignition (piloted or autoignition) is measured as a function of $\dot{q}''_e$ and m_{O_2}. From the least-squares analysis of the data using Eq. (8), $\dot{q}''_0$ and E_{eff} values are derived. The critical mass loss rate per unit sample surface area for ignition is also measured in the experiments.

3.3. *Data for Ignition Parameters*

The data for $\dot{q}''_0$ (also T_s), E_{eff}, $\dot{m}''_c$, and ϕ values for the polymers are listed in Table 1, where the data are taken from Ref. 4.

3.3.1. *Aliphatic-Type Polymers*

The data in Table 1 for aliphatic-type polymers show that (1) $\dot{q}''_0$ is of similar magnitude for granular polymers, (2) chlorinated polyethylenes and polyethylene foams have higher $\dot{q}''_0$ values than the granular polymers, (3) oxygenated polymers (polyoxymethylene and polymethylmethacrylate) have higher E_{eff} values than other polymers, and (4) foamed polyethylenes and polyethylenes with higher chlorine content have lower E_{eff} values than other polymers in Table 1. Polymers with low $\dot{q}''_0$ and E_{eff} values are expected to show easier ignition and faster surface flame spread.

The kinetic parameter ϕ is highest for polyoxymethylene, followed by polymethylmethacrylate and polyethylene. Polypropylene and polyethylene foams have similar ϕ values. The reactivities of the polymer vapors for oxygen follows the value of ϕ. Chlorine inhibits the reactivity of polyethylene vapors

TABLE 1
Ignition Parameters for Polymers[a]

Polymer	ρ (kg/m^3)	$\dot{q}''_0$ (kW/m^2)	T_s^b (K)	E_{eff} (kJ/m^2)	$\dot{m}''_c$ (g/m^2 sec)	ϕ
Aliphatic-type polymers						
Polyoxymethylene (granular)	1400	17	740	6137	4.5	0.43
Polymethylmethacrylate (granular)	1170	18	751	4207	4.4	0.28
Polyethylene (granular)	920	19	761	3744	2.5	0.27
Polypropylene (granular)	904	20	771	2895	2.7	0.24
Polyethylene foams (rigid)						
1	36	19	761	1776	2.6	0.24
2	56	24	807	1737	2.6	0.25
3	88	22	789	1429	2.5	0.25
4	143	22	789	1621	2.6	0.25
Chlorinated polyethylenes (granular)						
PE–25%Cl	1100	26	823	3011	6.6	0.12
PE–36%Cl	1160	26	823	1891	7.5	0.12
PE–42%Cl	–	22	789	1969	7.6	0.13
Polyvinyl chloride (granular)	1305	21	780	3320	–	–
Aromatic-type polymers						
Polystyrene (granular)	1051	29	846	6446	4.0	0.21
Epoxy–FR–fiberglass (laminate)	–	17	740	3950	–	–
Polystyrene foams (rigid)						
GM-47	16	27	831	2702	6.3	0.11
GM-49	34	18	751	3242	4.9	0.14
GM-51	34	19	761	2625	6.3	0.10
GM-53	29	21	780	2393	5.7	0.11
Polyurethane foams (flexible)						
GM-21	29	16	729	733	5.6	0.16
GM-23-FR	28	22	789	772	5.3	0.17
GM-25	44	30	852	695	5.7	0.15
GM-27-FR	43	30	852	695	6.5	0.12
I-A-$CaCO_3$	47	18	751	695	7.2	0.19
Polyurethane foams (rigid)						
GM-29	36	20	771	811	7.9	0.10
GM-31-FR	32	26	823	733	8.4	0.09
GM-37	321	28	838	1312	6.9	0.11
Polyisocyanurate foams (rigid)						
GM-41	33	23	798	926	6.8	0.12
GM-43	36	23	798	849	5.5	0.15
Phenolic foam (rigid)						
GM-57	67	45	944	386	5.5	0.17

[a]Data from Ref. 4.
[b]Assuming unit surface emissivity in Eq. (9).

for oxygen, where ϕ for the chlorinated polyethylenes is about half the value for polyethylene.

3.3.2. *Aromatic-Type Polymers*

In aromatic-type polymers, both $\dot{q}''_0$ and E_{eff} values for granular polystyrene are higher than the foams, i.e., foams are expected to show easier ignition and faster surface flame spread than the granular polymer. The kinetic parameter ϕ for the foams is about half the value for the granular polystyrene.

3.3.3. *Foamed Polymers*

In polyurethane foams, the flexible foams are expected to show easier ignition and faster surface flame spread than the rigid polyurethane and polyisocyanurate foams. The kinetic parameter for the rigid foams is smaller than the value for the flexible foams. Phenolic foam has the highest $\dot{q}''_0$ value, whereas its E_{eff} value is lowest.

From the data for foams in Table 1, (1) polyurethanes and polyisocyanurate foams are expected to show easier ignition and faster surface flame spread than the polyethylene foams, (2) polyethylene foams are expected to show easier ignition and faster surface flame spread than polystyrene foams, and (3) between foams and granular (or solid) polymers, foams are expected to show easier ignition and faster surface flame spread. Data in Table 1 show that fire retardants in the foams examined in our study do not appear to change the ignition and surface flame spread for the foams significantly. Inert surface coatings and/or other types of fire retardants may be beneficial in reducing the hazard due to ignition and surface flame spread.

From the data for granular polymers in Table 1, the polymers can be arranged in the following decreasing order of ignition and surface flame spread hazard: polypropylene, polyethylene, polymethylmethacrylate, polyoxymethylene, and polystyrene.

The polymers can be arranged in the following decreasing order of reactivity of polymer vapors for oxygen: polyoxymethylene ($\phi = 0.43$); polymethylmethacrylate, polyethylene, and polyethylene foams ($\phi \approx 0.25$); polystyrene ($\phi = 0.21$); flexible polyurethane foams and rigid phenolic foam ($\phi \approx 0.16$); chlorinated polyethylenes, rigid polyurethane and polyisocyanurate foams, and polystyrene foams ($\phi \approx 0.12$). The smaller the value of ϕ, the higher will be the yield of products of incomplete combustion.

3.4. *Conclusion*

The parameters $\dot{q}''_0$ (or surface temperature) and E_{eff} have been shown to be useful for evaluating ignition and presumably also for surface flame spread hazard of polymers under fixed experimental conditions. For reducing the hazard of foams, inert surface coatings and/or other types of fire retardants than used in the samples in our studies may be required. The reactivity of polymer vapors for oxygen has been expressed by a kinetic parameter ϕ. The yield of the products of incomplete combustion are expected to be an inverse function of ϕ.

4. *Mass Loss Rate in the Pyrolysis and Combustion of Polymers*

4.1. *Pyrolysis*

4.1.1. *Concept*

The pyrolysis of a polymer sample can be expressed in terms of a steady state heat balance at the surface,

$$m_p'' = \frac{\dot{q}_e''}{L_g} - \frac{\dot{q}_0''}{L_g} \tag{12}$$

where, m_p'' is the mass loss rate per unit sample surface area in pyrolysis (g/m^2 sec), L_g the "effective" heat of gasification of the polymer (kJ/g), $\dot{q}_e''$ the externally applied heat flux (kW/m^2), and $\dot{q}_0''$ is the "effective" surface reradiation loss (kW/m^2).

L_g includes quantities such as sensible heat of the polymers, heat from exo- or endothermic process in the condensed phase or at the surface, including surface char formation, heat of vaporization, heat of decomposition, heat of depolymerization, etc.

4.1.2. *Measurements and Calculations*

In the experiments the steady state value of $\dot{m}_p''$ is measured as a function of q_e''. From the least-squares analysis of the data using Eq. (12), L_g and $\dot{q}_0''$ values are derived.

4.1.3. *Data for "Effective" Heat of Gasification*

Table 2 lists the data for L_g. Some data for L_g measured in a differential scanning calorimeter (DSC) taken from Ref. 5 are also included in the table. The L_g value for distilled water calculated from its thermophysical properties is also included in the table. A comparison of the L_g values from DSC, the calculated L_g value for distilled water, and L_g values obtained from our apparatus shows excellent agreement and establishes the usefulness of our technique of obtaining L_g. The measuring of L_g using our technique is much

TABLE 2
"Effective" Heat of Gasification of Polymers

Polymer	Form	L_g (kJ/g) From our apparatus	L_g (kJ/g) From DSC[a]
Distilled water	Liquid	2.58	2.59[b]
Cellulosics			
Cellulose (Whatman filter paper	Sheets	3.55	
Corrugated paper	Sheets	2.21	
Douglas fir	Solid	1.82	
Plywood-FR	Sheets	0.95	
Polyethylenes (PE)			
PE-48%Cl	Granular	3.07	
PE-36%Cl	Granular	2.95	
PE (1)	Rigid foam	2.72	
PE (2)	Rigid foam	2.61	
PE (high density)	Granular	2.32	2.17
PE-25%Cl	Granular	2.12	
PE (3)	Rigid foam	1.79	
PE (low density)	Granular	1.75	
PE (4)	Rigid foam	1.50	
Polyvinyl chlorides (PVC)			
PVC-FR-plasticizer	Solid	2.52	
PVC	Granular	2.47	
PVC-plasticizer	Solid	1.68	
Polyolefins			
Polypropylene	Granular	2.03	2.00
Polyisoprene-65%Cl	Granular	2.02	
Polypropylene (sample from a commercial product)	Sheet	1.43	
Polyamide			
Nylon 6/6	Granular	2.35	
Polyoxide			
Polyoxymethylene	Granular	2.43	2.36
Acrylates			
Polymethylmethacrylate	Granular	1.63	1.61
Polystyrenes (PS)			
Acrylonitrile-Butadiene-Styrene (25-50-25%)	Granular	3.23	
Styrene-Butadiene (85-15%)	Granular	2.69	
PS (GM-47)	Rigid foam	1.94	
PS	Granular	1.70	1.79
PS (GM-51)	Rigid foam	1.39	
PS-FR	Rigid foam	1.36	1.38
PS (GM-49)	Rigid foam	1.32	
PW-FR (GM-53)	Rigid foam	1.31	
Polyurethanes (PU)			
PU-FR (GM-23)	Flexible foam	2.71	
PU-$CaCO_3$ (I-A)	Flexible foam	2.58	
PU (GM-21)	Flexible foam	1.96	
PU-FR (GM-27)	HR flexible foam	1.89	

TABLE 2 (Continued)

Polymer	Form	L_g (kJ/g) From our apparatus	From DSC[a]
PU	Flexible foam	1.23	1.17
PU (GM-25)	HR flexible foam	1.23	
PU-FR (GM-31)	Rigid foam	3.11	
PU (GM-29)	Rigid foam	3.11	
PU (GM-37)	Rigid foam	4.52	
Polyisocyanurates (PI)			
PI-FR-glass fiber	Rigid foam	3.74	
PI (GM-41)	Rigid foam	4.54	
PI (GM-43)	Rigid foam	4.54	
Phenolics			
Phenolic-FR	Rigid foam	3.74	
Phenolic	Solid	1.64	
Others			
Polycarbonate	Solid	2.07	
Polyester-FR-glass fiber	Composite	1.75	
Polyester-glass fiber	Composite	1.39	
Epoxy-FR-glass fiber	Composite	2.23	

[a]From Ref. 5.
[b]Calculated from literature data.

simpler than using the DSC (where several measurements are needed in order to calculate L_g).

L_g is one of the most important parameters used for evaluating the fire hazard of polymers. It is also useful in evaluating the effectiveness of additives such as fire retardants in the polymers, and other physical and chemical modifications of the polymers to reduce the fire hazard.

An examination of the data in Table 2 reveals the following information.

1. Polyisocyanurate foams, phenolic foam, and rigid polyurethane foam (GM-37), have highest L_g values in the table (these foams show surface charring).

2. Cellulose, polyethylene/36 and 48% Cl, acrylonitrile-butadiene-styrene, and rigid polyurethane foams also show higher L_g values, as well as surface charring.

3. The L_g value increases as the amount of chlorine in polyethylene is increased. The L_g value is lowered by branching and foaming the polyethylene.

4. Plasticizer lowers the L_g value of PVC. The fire retardant in the plasticized PVC appears to be effective in increasing the L_g value.

5. The thermal stability of polymers increases in the following order: polymethylmethacrylate, polystyrene, polypropylene, polyethylene (high density), polyoxymethylene, polyvinyl chloride, and cellulose and is consistent with Madorsky's prediction.[6]

6. Acrylonitrile and butadiene as copolymers in polystyrene increase the L_g values. Foamed polystyrene has a lower L_g value than the granular polymer. Fire retardants in the foam samples of polystyrene do not appear to have significant effect on L_g.

7. High-resilience flexible polyurethane foams have lower L_g values than other flexible foams. The rigid polyurethane foams have higher L_g values than the flexible foams. Fire retardants in flexible polyurethane foams increase the L_g values of the foams.

4.2. *Combustion*

4.2.1. *Concept*

The combustion of a polymer sample can be expressed in terms of a steady state heat balance at the surface:

$$\dot{m}''_b = \frac{\dot{q}''_e}{L_g} + \frac{\dot{q}''_{fr}}{L_g} + \frac{\dot{q}''_{fc}}{L_g} - \frac{\dot{q}''_0}{L_g} \tag{13}$$

where $\dot{m}''_b$ is the mass loss rate per unit sample surface area in combustion (g/m^2 sec) $\dot{q}''_e$ the externally applied heat flux (kW/m^2), L_g the "effective" heat of gasification of the polymer (kJ/g), $\dot{q}''_{fr}$ the flame radiative heat flux to the surface (kW/m^2), $\dot{q}''_{fc}$ the flame convective heat flux to the surface (kW/m^2), and $\dot{q}''_0$ is the "effective" surface reradiation loss (kW/m^2).

In the absence of $\dot{q}''_e$ under normal conditions for small samples, $\dot{q}''_{fc} \gg \dot{q}''_{fr}$ and the combustion is defined as dominated by convective heat transfer. For large samples, as in real fires, in the absence of q''_e under normal conditions, $\dot{q}''_{fr} \gg \dot{q}''_{fc}$ and the combustion is defined as dominated by radiative heat transfer. Since the purpose of laboratory scale experiments, where small samples are used, is to predict the behavior of polymers in large-scale or real fires, it is essential that the condition where $\dot{q}''_{fr} \gg \dot{q}''_{fc}$ be established prior to any other measurements of flammability properties. For laboratory scale samples this condition can be achieved by increasing the value of $\dot{q}''_e$ and m_{O_2}. An increase in $\dot{q}''_e$ probably increases the residence time of the fuel vapors in the flame and preflame zone because of the different dynamics of the larger fire plume. An increase in the residence time of the fuel vapors is expected to increase the degradation of the fuel vapors. An increase in m_{O_2} is expected to increase the flame temperature, which is helpful in the further degradation of the fuel vapors. The net effect of the combination of $\dot{q}''_e$ and m_{O_2} is an increase in flame radiation. In the laboratory scale experiments it is thus possible to examine the combustion of polymers over a wide range of experimental conditions where the combustion can be changed from a process dominated by flame convection to one dominated by flame radiation.

For flame-convection-dominated combustion of fuels, the mass loss rate of the fuel can be expressed as[7]

$$\dot{m}''_b = \frac{h}{C_p} \ln (1 + B_{\text{conv}}) \tag{14}$$

where, $\dot{m}''_b$ is the mass loss rate per unit sample surface area in the combustion of the fuel (g/m^2 sec), h the convective heat transfer coefficient (kW/m^2 K), C_p the specific heat of air (kJ/g K), and B_{conv} is the transfer number given by the following equation:

$$B_{\text{conv}} = \frac{m_{O_2} H_c/k_{O_2} - C_p (T_s - T_a)}{L_g} \tag{15}$$

where m_{O_2} is the mass fraction of oxygen in the environment, H_c the net heat of complete combustion of the fuel (kJ/g), k_{O_2} the theoretical stoichiometric mass of oxygen consumed per unit mass of fuel for complete combustion, T_s the surface temperature of the fuel (K), T_a the ambient temperature of the fuel (K), L_g is the effective heat of gasification of the fuel (kJ/g).

If the combustion is modified by external heat flux and flame radiation, Eq. (15) is modified[8,9] as

$$B_{\text{rad}} = \frac{m_{O_2} (H_c/k_{O_2}) (\chi_{\text{conv}}/\chi_{\text{actual}}) - C_p (T_s - T_a)}{L_g (1 - E)} \tag{16}$$

where B_{rad} is the radiation-modified transfer number, χ_{conv} the convective fraction of the theoretical stoichiometric heat release rate for complete combustion χ_{actual} the actual fraction of the theoretical stoichiometric heat release rate for complete combustion, and E is given[8] as

$$E = \frac{\dot{q}''_e + \dot{q}''_{\text{fr}} - \dot{q}''_0}{\dot{m}''_b \cdot L_g} \tag{17}$$

Equations (14)–(17) when combined with Eq. (13) are very useful for estimating the value of $\dot{q}''_{\text{fc}}$, which can be expressed as

$$\frac{\dot{q}''_{\text{fc}}}{L_g} = \frac{[m_{O_2} (H_c/k_{O_2}) (\chi_{\text{conv}}/\chi_{\text{actual}}) - C_p (T_s - T_a)] (\dot{m}''_b/L_g)}{\exp (\dot{m}''_b C_p/h) - 1} \tag{18}$$

By substituting all the known values of the quantities in Eq. (13), $\dot{q}''_{\text{fr}}/L_g$ thus can be estimated.

In the experiments $\dot{q}''_{\text{fr}}/L_g$ and $\dot{q}''_{\text{fc}}/L_g$ are estimated under various combinations of $\dot{q}''_e$ and m_{O_2}, to identify the combination where $\dot{q}''_{\text{fr}} \gg \dot{q}''_{\text{fc}}$. Once this combination has been identified, the value of $\dot{q}''_e$ is maintained constant and the m_{O_2} value is increased to examine the flame-radiative characteristic of the fuels and to obtain the asymptotic value of $\dot{q}''_{\text{fr}}/L_g$. $\dot{q}''_{\text{fr}}/L_g$ is one of the most important parameters expressing the fire hazard of various fuels and can be used to classify the polymers. The asymptotic value of $\dot{q}''_{\text{fr}}/L_g$

is expected to represent the value for the combustion of large samples of the polymers. Another useful quantity that can be derived is the magnitude of change of $\dot{q}''_{fr}/L_g$ for a fixed change in m_{O_2} values for the polymer, i.e., $\Delta(\dot{q}''_{fr}/L_g)/\Delta m_{O_2}$. The higher the value of $\Delta(\dot{q}''_{fr}/L_g)/\Delta m_{O_2}$, the greater is the expected fire hazard of the polymers. In addition, the magnitudes of $\dot{q}''_e$ and m_{O_2} are important for changing combustion from a process dominated by flame convection to one dominated by flame radiation. The smaller the magnitudes of q''_e and m_{O_2} required for the change, the greater is the expected fire hazard of the polymers.

4.2.2. *Measurements and Calculations*

In the experiments, the apparatus is calibrated for the h/C_p value by using methanol. The value of $\dot{m}''_b$ for methanol is measured in the non-luminous flame zone at various $\dot{q}''_e$ and m_{O_2} values. Equation (14) is then fused to calculate h/C_p, where B_{conv} for methanol is calculated from its well-known thermophysical properties.

For the experimental conditions in our apparatus, $h/C_p = 13.61$ g/m^2 sec.[9]

Values of $\dot{m}''_b$ for various fuels are then measured under conditions identical to those used for methanol for a variety of combinations of $\dot{q}''_e$ and m_{O_2}. Values of L_g and $\dot{q}''_0$ (and T_s) (see Tables 1 and 2) measured in the apparatus are used. The heat of combustion of the fuels is measured in an oxygen bomb calorimeter and is corrected for water as a gas. The elemental compositions of the fuels are measured by using a microanalytical technique from which k_{O_2} is calculated. In the experiments χ_{conv} and χ_{actual} are also measured for each combination of $\dot{q}''_e$ and m_{O_2} values (described in a later section). The C_p value used is for air at ambient temperature. Equations (18) and (13) are then used to calculate $\dot{q}''_{fc}/L_g$ and $\dot{q}''_{fr}/L_g$, respectively.

4.2.3. *Data for $\dot{q}''_{fc}/L_g$ and $\dot{q}''_{fr}/L_g$*

The data for $\dot{q}''_{fc}/L_g$ and $\dot{q}''_{fr}/L_g$ for various liquids and polymers are listed in Table 3. The data in Table 3 for higher $\dot{q}''_e$ values show that, for most polymers, $\dot{q}''_{fr} \gg \dot{q}''_{fc}$; thus one of the most important conditions for the prediction of fire hazard for large-scale or real fires has been satisfied. The variation of $\dot{q}''_{fr}/L_g$ with m_{O_2} when $\dot{q}''_{fr} \gg \dot{q}''_{fc}$ shows the flame-radiative characteristic of the fuel. Figure 2 is a plot of $\dot{q}''_{fr}/L_g$ as a function of m_{O_2}. In the linear portions of the curves, $\Delta(\dot{q}''_{fr}/L_g)/\Delta m_{O_2} \approx 70$, 120, 130, 150, and 244 g/m^2 sec for methanol, polymethylmethacrylate, polyethylene, flexible polyurethane foams GM-21 and GM-25, respectively. $\Delta(\dot{q}''_{fr}/L_g)/\Delta m_{O_2}$ is related to the ease with which flame radiation can be achieved; in that respect, an

TABLE 3
Ratio of Flame-Convective and Flame-Radiative Heat Flux to the Surface-to-Heat of Gasification for Various Liquids and Polymers

Liquid–polymer	m_{O_2}	$\dot{q}''_e$ (kW/m^2)	$\dot{q}''_{fc}/L_g$ (g/m^2 sec)	$\dot{q}''_{fr}/L_g$ (g/m^2 sec)
Methanol (liquid)	0.233	0	15.5	1.7
	0.233	10.0	10.7	4.3
	0.233	20.0	8.6	3.1
	0.233	30.0	5.4	5.2
	0.172	52.0	2.1	0
	0.233	52.0	2.0	7.8
	0.270	52.0	1.4	13.2
	0.301	52.0	1.8	12.0
	0.371	52.0	1.8	16.2
	0.446	52.0	1.7	19.8
	0.518	52.0	1.7	20.9
	0.233	55.3	1.7	8.5
Acetone (liquid)	0.233	0	23.3	0
Heptane (liquid)	0.233	0	13.3	16.4
Aniline (liquid)	0.233	0	12.0	11.4
	0.233	20.0	1.9	29.0
Benzene (liquid)	0.233	0	1.8	55.8
Cellulosics				
Cellulose (Whatman filter paper)	0.233	52.4	1.9	7.8
Polyethylenes (PE)				
PE, low density (granular)	0.233	0	11.5	2.6
	0.150	38.9	0.6	8.6
	0.168	38.9	0.8	10.5
PE, low density (granular)	0.190	38.9	1.0	12.6
	0.208	38.9	1.0	14.9
	0.233	38.9	1.1	17.3
	0.283	38.9	1.1	24.7
	0.310	38.9	1.0	30.1
	0.233	45.2	1.0	17.4
	0.233	52.3	0.8	17.7
PE–25%Cl (granular)	0.233	38.9	1.4	13.1
PE (rigid foam 1)	0.233	26.5	5.4	0.1
	0.233	38.9	3.3	0
	0.233	52.4	1.9	3.3
PE (rigid foam 2)	0.233	26.5	4.5	2.2
	0.233	38.9	2.9	3.0
	0.233	52.4	2.0	4.7
PE (rigid foam 3)	0.233	26.5	8.4	0
	0.233	38.9	3.7	4.1
	0.233	52.4	3.1	2.3
PE (rigid foam 4)	0.233	26.5	10.7	0
	0.233	38.9	4.5	2.9
	0.233	52.4	3.0	4.1

TABLE 3 (Continued)

Liquid-polymer	m_{O_2}	$\dot{q}''_e$ (kW/m^2)	$\dot{q}''_{fc}/L_g$ (g/m^2 sec)	$\dot{q}''_{fr}/L_g$ (g/m^2 sec)
Polyvinyl chloride (PVC)				
PVC (granular)	0.233	52.4	1.5	10.4
Polyolefin				
Polypropylene (granular)	0.233	0	8.8	7.3
	0.233	52.4	1.2	27.9
	0.259	52.4	1.2	30.6
	0.281	52.4	1.4	29.6
	0.313	52.4	1.4	31.8
Polyamide				
Nylon 6/6 (granular)	0.233	52.4	0.8	18.1
Polyoxide				
Polyoxymethylene (granular)	0.233	0	11.6	3.3
	0.233	52.4	4.3	3.6
Acrylate				
Polymethylmethacrylate (granular)	0.233	0	10.8	8.5
	0.174	52.4	2.1	8.6
	0.205	52.4	2.3	10.7
	0.233	52.4	2.0	16.6
	0.251	52.4	2.1	17.3
	0.279	52.4	2.3	18.5
	0.317	52.4	1.7	28.0
	0.401	52.4	1.5	34.7
Polystyrene (PS)				
Polystyrene (granular)	0.233	0	10.0	14.4
	0.233	32.5	1.0	36.3
	0.164	38.9	0.5	27.2
	0.233	38.9	0.8	34.4
	0.262	38.9	1.0	32.5
Polyurethanes (PU)				
Flexible foam-$CaCO_3$ (I-A)	0.233	26.4	3.6	3.6
	0.233	38.9	1.7	6.7
	0.233	45.2	0.9	8.9
	0.233	52.4	1.1	8.0
	0.233	59.8	0.8	9.9
Flexible foam (GM-21)	0.160	45.2	1.5	4.2
	0.180	45.2	1.3	6.9
	0.213	45.2	1.1	13.1
	0.233	45.2	1.0	14.8
Flexible foam-FR (GM-23)	0.167	45.2	0.7	12.0
	0.233	45.2	0.5	19.3
	0.273	45.2	0.5	18.8
HR flexible foam (GM-25)	0.207	26.4	3.2	18.8
	0.214	26.4	2.9	20.4
	0.233	26.4	1.8	26.0
	0.301	26.4	0.7	41.7
	0.233	31.8	0.7	31.5

TABLE 3 (Continued)

Liquid-polymer	m_{O_2}	$\dot{q}''_e$ (kW/m²)	$\dot{q}''_{fc}/L_g$ (g/m² sec)	$\dot{q}''_{fr}/L_g$ (g/m² sec)
	0.233	38.9	0.7	30.3
	0.190	52.4	0.2	17.6
	0.201	52.4	0.3	20.2
	0.217	52.4	0.3	23.9
	0.233	52.4	0.3	30.0
	0.255	52.4	0.3	32.7
	0.233	59.8	0.4	28.6
HR Flexible foam-FR (GM-27)	0.233	31.8	0.9	26.3
	0.233	38.9	0.8	27.1
	0.233	52.4	0.4	28.6
Others				
Epoxy-FR-glass fibers	0.233	52.4	0.5	9.8
	0.279	52.4	0.1	14.8
	0.334	52.4	0.2	14.8

increase in $\Delta\dot{q}''_{fr}/L_g)/\Delta m_{O_2}$ represents an increase in fire hazard. In Fig. 2, $\dot{q}''_{fr}/L_g$ for methanol appears to reach an asymptotic value of 21 g/m² sec. For the combustion of methanol in large pools in normal air, the measured values of $\dot{m}''_g$ have been reported to be 20.5, 19.5, and 20.8 g/m² sec for pool areas of 1.17, 2.37, and 4.67 m², respectively[10] and are very close to our estimated value of 21 g/m² sec based on $\dot{q}''_{fr}/L_g$. None of the values of $\dot{q}''_{fr}/L_g$ for polymers in Fig. 2 appear to reach their asymptotic values. We are, thus, extending the

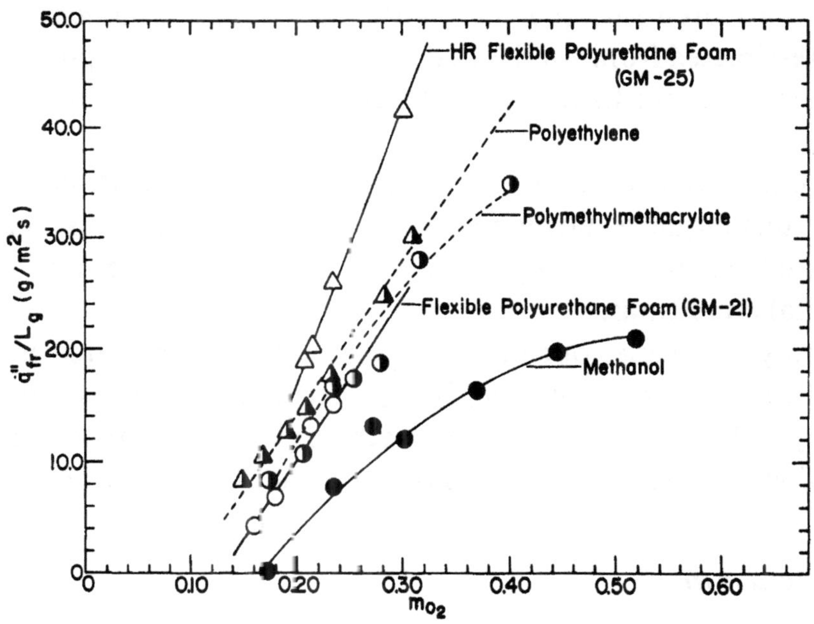

FIGURE 2. $\dot{q}''_{fr}/L_g$ as a function of m_{O_2} ($\dot{q}''_{fc}$ is negligibly small for all the samples).

TABLE 4
Comparative $\dot{q}''_{fr}$ Values for Polymers

Polymer	$\dot{q}''_{fr}$ (kW/m^2) de Ris[a]	$\dot{q}''_{fr}$ (kW/m^2) Ours[b]
Polymethylmethacrylate	36.5[c]	34.7[d, e]
Polypropylene	37.7	61[d]
Polystyrene	54.8	59.3[d]

[a]Combustion in normal air, no external heat flux, sample surface area $\approx$0.093 m^2. Samples in pool configuration. (11)
[b]Sample surface area $\approx$0.008 m^2; samples in pool configuration.
[c]For wall fire at a height of 3.56 m.
[d]Asymptotic values of $\dot{q}''_{fr}$ obtained by increasing m_{O_2} values with $\dot{q}''_e$ value. Maintained constant such that $\dot{q}''_{fc}$ is a negligibly small quantity and is independent of m_{O_2} values.
[e]$\dot{q}''_{fr}$ is close to the asymptotic value.

range of m_{O_2} values for all the polymers and liquids listed in Table 3 such that $\dot{m}''_b$ values can be estimated for large-size samples.

The following equation has been proposed by de Ris[11] for the estimation of $\dot{q}''_{fr}$ where the principles used are entirely different than those used by us:

$$\dot{q}''_{fr} = \sigma T_f^4 (1 - e^{-Kl_m}) \tag{19}$$

where, σ is the Stefan–Boltzmann constant (kW/m^2 K^4), T_f the flame radiation temperature (K), K the absorption-emission coefficient (m^{-1}), and l_m is the mean beam length (m). For pool fires 0.093 m^2 in surface area, T_f is measured quite reliably with the Schmidt technique. K is obtained from the emissivity measurement, together with an average turbulent flame-path length (l_m) estimated from a series of digitized flame photographs. Table 4 lists the $\dot{q}''_{fr}$ values estimated from Eq. (19) by de Ris[11] and some of our values from Table 3.

The comparative data in Table 4 show that the estimation of $\dot{q}''_{fr}$ values using our technique is fairly reasonable. Our technique is very simple: Samples used are small; and $\dot{q}''_{fr}$ values can be estimated over a wide range of experimental conditions.

For polypropylene, our data suggest that if sample area is increased beyond 0.093 m^2 for pool fires, $\dot{q}''_{fr}$ is expected to increase beyond the values reported by de Ris.[11] For polymethylmethacrylate, our data in Fig. 2 suggest that the $\dot{q}''_{fr}$ value is expected to be slightly higher than reported in Table 4.

4.3. *Conclusions*

For evaluating the fire hazard of polymers the parameter $\dot{q}''_{fr} / L_g$ has been shown to be useful. For obtaining generalized flame radiative behavior of polymers, the $\dot{q}''_{fr}/L_g$ value is examined under increasing m_{O_2} values, when $\dot{q}''_{fr} \gg \dot{q}''_{fc}$. The polymers may be classified for their fire hazard on the basis of the asymptotic value of $\dot{q}''_{fr}/L_g$. The relative magnitudes of $\Delta(\dot{q}''_{fr}/L_g)/\Delta m_{O_2}$ for polymers are also useful for evaluating the fire hazard related to the ease with which flame radiation is achieved.

5. *Mass Generation (or Depletion) Rates of Products*

5.1. *Concept*

In the experiments the mass generation rates of CO, CO_2, and total gaseous hydrocarbons (as CH_4) and mass depletion rate of O_2 are measured. The mass generation (or depletion) rates can be used for calculating a carbon balance. Using polymethylmethacrylate (PMMA) as an example, we can write

$$\mathrm{nC_5H_8O_2 + an\ O_2 \longrightarrow nbCO_2 + ncCO + ndHC + nf\ pyrolyzate} \quad (20)$$

where HC represents the total gaseous hydrocarbons (measured as CH_4), and pyrolyzate is the mixture of carbon-containing compounds except CO, CO_2, and HC. In terms of carbon, Eq. (20) can be expressed as follows, for the conversion of PMMA to products:

$$\frac{60}{100} \times \dot{m}''_b \equiv \frac{12}{44} \cdot \dot{M}''_{CO_2} + \frac{12}{28} \cdot \dot{M}''_{CO} + \frac{12}{16} \cdot \dot{M}''_{HC} + X \cdot \dot{M}''_{pyr} \quad (21)$$

where HC is assumed to CH_4 and X is an unknown coefficient.

Rearranging Eq. (21) gives

$$\frac{\dot{M}''_{CO_2}}{2.2\dot{m}''_b} + \frac{\dot{M}''_{CO}}{1.4\dot{m}''_b} + \frac{\dot{M}''_{HC}}{0.8\dot{m}''_b} + \frac{X\dot{M}''_{Pyr}}{0.6\dot{m}''_b} = 1.0 \quad (22)$$

or

$$\frac{\dot{M}''_{CO_2}}{k_{CO_2} \cdot \dot{m}''_b} + \frac{\dot{M}''_{CO}}{k_{CO} \cdot \dot{m}''_b} + \frac{\dot{M}''_{HC}}{k_{HC} \cdot \dot{m}''_b} + \frac{\dot{M}''_{pyr}}{k_{pyr} \cdot \dot{m}''_b} = 1.0 \quad (23)$$

or

$$f_{CO_2} + f_{CO} + f_{HC} + f_{pyr} = 1.0 \quad (24)$$

In Eq. (23) k_i is the theoretical stoichiometric mass of the product produced per unit mass of the fuel converted completely to the product. For example, $k_{CO_2} = (5 \times 44)/100 = 2.2$. In Eq. (24) f_i is defined as the fractional theoretical stoichiometric yield of product or the fraction of carbon in the fuel converted to the product.

In Eq. (23), $\dot{M}''_{pyr}$ and k_{pyr} are unknown and, thus, f_{pyr} cannot be calculated; f_{pyr}, however, can be calculated from Eq. (24), i.e.,

$$f_{pyr} = 1 - (f_{CO_2} + f_{CO} + f_{HC}) \tag{25}$$

In Eq. (25) f_{pyr} can be expressed as follows:

$$f_{pyr} = (f_{pyr})_{liquid\text{-}1} + [(f_{pyr})_{liquid\text{-}2} + (f_{pyr})_{solid}] \tag{26}$$

where liquid-1 represents high-vapor-pressure liquid compounds in the pyrolyzate and liquid-2 the low-vapor-pressure liquid compounds in the pyrolyzate.

In our experiments, the pyrolyzate fraction collected on the filter paper (espressed as pyr-f) represents liquid-2 and solid in Eq. (26). If we assume that the pyrolyzate fraction collected on the filter paper is predominantly carbon, then from the measured values of $\dot{M}''_{pyr,f}$, $\dot{m}''_b$, and $k_{pyr,f}$, $(f_{pyr})_{liquid\text{-}2} + (f_{pyr})_{solid}$ can be calculated. The difference between f_{pyr} and $(f_{pyr})_{liquid\text{-}2} + (f_{pyr})_{solid}$, thus, is the pyrolyzate fraction which cannot be collected on the filter paper and is equal to $(f_{pyr})_{liquid\text{-}1}$.

By heating the pyrolyzate fraction collected on the filter paper to 373 K in vacuum for 24 hr, liquid-2 is eliminated and thus $(f_{pyr})_{solid}$ can be calculated.

Values of f_i are among the most important parameters used for evaluating the fire hazard of polymers in terms of actual heat release rates, generation rates of toxic compounds, and optical transmission (or visibility) through the products.

5.2. *Measurements and Calculations*

For a variety of combinations of $\dot{q}''_e$ and m''_{O_2} values, $\dot{M}''_{CO_2}$, $\dot{M}''_{CO}$, M''_{HC} and $\dot{M}''_{O_2}$ are obtained from Eq. (3). $\dot{M}''_{pyr\text{-}f}$ (pyrolyzate fraction collected on the filter paper) is calculated from Eq. (4). $(\dot{M}''_{pyr})_{solid}$ (solid compounds in the pyrolyzate fraction collected on the filter) is also calculated from Eq. (4) after heating the filter paper with pyrolyzate fraction at 373 K under vacuum for 24 hr. Values of $\dot{m}''_b$ are also measured. The elemental compositions of the polymers are measured using a microanalytical technique from which k_{CO_2}, k_{CO}, k_{HC}, and $k_{pyr,f}$ are calculated (pyr-f is the pyrolyzate fraction collected on the filter paper and is assumed to be predominantly carbon). Values of f_i are then calculated from Eq. (23). $(f_{pyr})_{liquid\text{-}1}$ is calculated from Eqs. (25) and (26) and represents the carbon fraction which cannot be collected on the filter paper because of the high vapor pressure of the liquid compounds in the products.

5.3. *Data for the Distribution of Carbon in the Combustion Products*

Data for the distribution of carbon in the combustion products of liquids and polymers are listed in Table 5. The data indicate the following.

TABLE 5
Distribution of Carbon in the Combustion Products of Liquids and Polymers[a]

Liquid–polymer	$\dot{q}''_e$ (kW/m^2)	m_{O_2}	Gases, $f_i \times 100$			Pyrolyzate, $f_i \times 100$			
			CO_2	CO	HC	Pyr[b]	Pyr-f[c]	$(Pyr)_{solid}$[d]	$(Pyr)_{liquid\text{-}1}$[e]
Aliphatic-type fuels									
Liquids									
Methanol	0	0.233	99.3	0.1	0	0.6	0	–	0.6
Acetone	0	0.233	76.2	0.1	0	23.7	0	–	23.7
Heptane	0	0.233	66.8	3.8	5.7	23.6	–	–	–
Polymers									
Cellulose	52.4	0.233	71.6	0.5	0.1	27.8	–	–	–
Polyoxymethylene	0	0.233	75.4	0.2	0.2	24.2	–	–	–
	38.9	0.233	77.7	0.1	0.1	22.1	–	–	–
	52.4	0.233	76.3	0.2	0.1	23.4	–	–	–
Polymethylmethacrylate	0	0.233	86.6	0.4	0	13.0	2.2	1.2	10.8
	4.0	0.233	78.9	0.5	0	20.6	3.2	1.5	17.4
	9.6	0.233	87.3	0	0	12.7	2.8	–	9.9
	11.3	0.288	96.3	0.3	0	3.4	3.4	–	0
	16.4	0.139	70.8	0.7	0	28.5	0.5	–	28.0
		0.154	88.5	1.0	0	10.5	4.3	–	6.2
		0.169	85.0	1.0	0	14.0	3.4	2.2	10.6
		0.184	76.8	1.1	0	22.1	4.8	–	17.3
		0.233	84.9	0.3	0	14.8	5.6	–	9.2
		0.236	80.7	0.7	0	18.6	3.3	–	15.3
		0.244	80.9	0.7	0	18.4	4.7	3.3	13.7
		0.291	78.8	0.4	0	20.8	2.3	–	18.5
	24.6	0.233	81.5	0.3	0	18.2	3.7	3.6	14.5
		0.244	78.2	0.9	0	20.9	5.5	4.4	15.4
	26.9	0.233	78.8	0.8	0.1	20.3	4.2	3.1	16.1
		0.284	92.2	0.4	0	7.4	4.3	–	3.1

TABLE 5 (Continued)

Liquid–polymer	$\dot{q}''_e$ (kW/m^2)	m_{O_2}	Gases, $f_i \times 100$			Pyrolyzate, $f_i \times 100$			
			CO_2	CO	HC	Pyr[b]	Pyr-f[c]	$(Pyr)_{solid}$[d]	$(Pyr)_{liquid\text{-}1}$[e]
	27.8	0.233	75.3	0.8	0.1	23.8	4.2	3.7	19.6
	29.0	0.233	76.5	1.1	0.1	22.3	4.7	4.4	17.6
	30.4	0.154	90.0	0.8	0	9.2	2.0	–	7.2
		0.233	73.8	1.6	0.2	24.4	5.2	3.6	19.2
	32.7	0.233	76.4	1.1	0.2	22.3	6.3	3.8	16.0
	39.7	0.233	69.6	1.0	0.2	29.2	5.2	–	24.0
	52.4	0.233	70.0	1.5	0.2	28.3	5.5	–	22.8
Polypropylene	0	0.233	74.0	1.4	0.2	24.4	13.0	–	11.4
	39.7	0.233	58.0	1.3	2.0	38.7	–	–	–
	52.4	0.233	61.2	2.7	4.6	31.5	–	–	–
Nylon 6/6	52.4	0.233	71.5	2.6	2.1	23.8	21.8	–	–
Polyethylenes (PE)									
PE	0	0.233	72.2	1.0	0	26.8	–	–	–
	21.0	0.233	66.0	1.8	1.5	30.7	–	–	–
	40.2	0.233	72.5	2.7	3.6	21.2	–	–	–
PE foams									
1	38.9	0.233	68.9	1.1	0.3	29.7	10.4	–	19.3
	52.0	0.233	67.1	1.9	0.5	30.5	–	–	–
2	26.5	0.233	78.1	1.3	0.9	19.7	20.1	–	0
	38.9	0.233	70.5	1.8	0.4	27.3	10.9	–	16.4
	52.0	0.233	68.0	2.9	1.1	28.0	18.0	–	10.0
3	26.5	0.233	72.3	1.3	0.2	26.2	16.5	–	9.7
	38.9	0.233	68.5	2.4	1.1	28.0	10.4	–	17.6
	52.0	0.233	62.2	3.7	2.5	31.6	9.5	–	22.1
4	26.5	0.233	73.3	1.4	0.3	25.0	11.9	–	13.1
	38.9	0.233	68.0	1.6	0.7	29.7	8.5	–	21.2
	52.0	0.233	51.2	3.7	1.9	43.2	13.7	–	29.5

PE–25%Cl	35.1	0.233	59.7	3.4	1.8	35.1	–	–	–
	38.9	0.233	54.4	4.2	2.9	38.5	39.3	–	0
PE–36%Cl	26.8	0.233	31.6	3.9	2.7	61.8	63.6	–	0
PE–48%Cl	20.1	0.233	26.8	4.6	2.9	65.7	–	–	–
Polyvinylchloride	52.4	0.233	32.7	6.3	5.5	55.5	55.5	–	0
Aromatic-type fuels									
Liquids									
Aniline	0	0.233	62.9	2.1	0.5	34.5	–	–	–
	20.0	0.233	46.3	3.3	1.8	48.6	–	–	
Styrene	0	0.233	54.0	3.0	2.2	40.8	39.7	–	1.1
Benzene	0	0.233	47.1	5.4	3.3	44.2	–	–	–
Polymers									
Styrene–butadicne	0	0.233	52.3	2.6	1.2	43.9	25.4	–	18.5
	45.2	0.233	47.1	2.2	1.0	49.7	34.4	–	15.3
Polystyrene	0	0.233	56.7	4.0	1.3	38.0	13.8	11.3	24.2
	3.0	0.233	58.2	2.9	2.3	36.6	19.4	–	17.2
	5.4	0.233	55.6	3.5	2.7	38.2	14.8	12.7	23.4
	8.1	0.233	54.6	2.7	2.3	40.4	19.3	14.0	21.1
		0.262	59.1	3.2	1.0	36.7	12.6	11.3	24.1
		0.281	56.5	2.1	1.3	40.1	15.3	13.4	24.9
		0.321	59.1	1.4	0.8	38.7	9.9	9.0	28.8
	11.3	0.233	52.4	3.3	3.1	41.2	16.9	15.7	24.3
	14.7	0.233	54.9	2.9	3.2	39.0	18.5	16.9	20.5
	16.4	0.233	55.2	3.2	2.6	39.0	15.6	13.8	23.4
	22.6	0.233	51.6	4.0	5.5	38.9	15.9	13.7	23.0
	24.6	0.233	45.9	2.9	5.0	46.2	19.9	–	26.3
	30.4	0.233	49.4	3.0	3.3	44.3	17.0	–	27.3
	32.5	0.233	36.6	2.6	3.8	57.0	–	–	–
	38.9	0.233	38.6	2.7	3.0	55.7	–	–	–
	39.7	0.233	43.7	2.7	3.8	49.8	–	–	–

TABLE 5 (Continued)

Liquid–polymer	$\dot{q}''_e$ (kW/m^2)	m_{O_2}	Gases, $f_i \times 100$			Pyrolyzate, $f_i \times 100$			
			CO_2	CO	HC	Pyr[b]	Pyr-f[c]	$(Pyr)_{solid}$[d]	$(Pyr)_{liquid\text{-}1}$[e]
Epoxy–FR-fiberglass	38.9	0.233	47.9	2.6	0.6	48.9	12.9	–	36.0
	47.0	0.233	43.9	2.4	0.7	53.0	14.6	–	38.4
	52.4	0.233	49.0	2.8	0.9	47.3	13.1	–	34.2
		0.279	46.1	2.2	0.6	51.1	14.4	–	36.7
		0.334	45.3	1.5	0.3	52.9	12.1	–	40.8
	59.7	0.233	40.4	2.2	0.5	56.9	14.2	–	42.7
Polystyrene foam (GM-47)	0	0.233	48.2	2.7	1.0	48.1	26.8	–	21.3
Polyurethane foams									
I-A–$CaCO_3$	0	0.233	68.3	0.3	0.04	31.4	29.3	–	2.1
	10.0	0.233	69.5	0.5	0.03	30.0	23.1	–	6.9
	20.1	0.233	63.8	0.4	0.1	35.7	33.2	–	2.5
	26.4	0.233	61.8	0.5	0.1	37.6	36.3	–	1.3
	38.9	0.233	67.1	0.7	0.1	32.1	31.2	–	0.9
	45.2	0.233	64.8	0.8	0.1	34.3	25.9	–	8.4
	52.3	0.233	63.9	0.7	0.1	35.3	26.8	–	8.5
	59.8	0.233	65.9	1.2	0.2	32.7	33.8	–	0
GM-21	45.2	0.160	61.3	0.7	0.8	37.2	32.0	–	5.2
		0.181	62.7	0.9	0.2	36.2	24.5	–	11.7
		0.212	60.4	0.7	0.2	38.7	33.1	–	5.6
		0.233	58.6	0.5	0.2	40.7	19.3	–	21.4
GM-23–FR	45.2	0.168	57.4	3.3	1.1	38.2	36.8	–	1.4
		0.177	57.8	2.5	0.7	39.0	–	–	–
		0.207	57.0	2.0	0.5	40.5	31.5	–	9.0
		0.233	60.2	1.9	0.5	37.4	30.5	–	6.9
		0.273	62.7	1.0	0.1	36.2	–	–	–

GM-25	26.4	0.207	60.6	2.4	0.5	36.5	–	–	–
		0.214	64.3	2.6	0.6	32.5	30.8	–	1.7
		0.233	63.0	1.9	0.3	34.8	32.5	–	2.3
		0.301	60.3	1.2	0.2	38.3	27.2	–	11.1
	31.8	0.233	57.5	1.1	0.8	40.6	–	–	–
	38.9	0.233	56.6	2.8	0.9	39.7	26.2	–	13.5
	52.3	0.233	56.9	2.5	1.6	39.0	30.9	–	8.1
	59.8	0.233	58.2	3.6	1.9	36.3	38.5	–	0
GM-27-FR	31.8	0.233	53.2	3.6	1.3	41.9	36.6	–	5.3
	38.9	0.233	51.2	3.8	1.3	43.7	–	–	–
	52.3	0.233	50.8	4.5	2.2	42.5	–	–	–
GM-29	52.0	0.233	40.1	5.4	1.5	53.0	32.6	–	20.4
GM-31-FR	52.0	0.233	40.7	5.1	1.9	52.3	51.2	–	1.1

[a] Fires not fully ventilated.
[b] Total pyrolyzate calculated from Eq. (25).
[c] Pyr-*f* pyrolyzate fraction collected on the filter paper consisting of low-vapor-pressure liquids and solid compounds.
[d] $(Pyr)_{solid}$: solid compounds in the pyrolyzate collected on the filter paper.
[e] $(Pyr)_{liquid\text{-}1}$: high-vapor-pressure liquids in the pyrolyzate which cannot be collected on the filter paper calculated from Eqs. (25) and (26).

1. A major portion of the carbon in the fuel is converted to CO_2 and pyrolyzate.

2. For most polymers, all the pyrolyzate cannot be collected on the filter paper, because of the production of substantial amounts of high-vapor liquid compounds.

3. The carbon in the fuel is distributed as follows in the pyrolyzate: (a) polymethylmethacrylate: ~20% in total pyrolyzate, ~3% in solid compounds in the pyrolyzate, ~1% in low-vapor-pressure liquid compounds in the pyrolyzate, and ~ 16% in high-vapor-pressure liquid compounds in the pyrolyzate; (b) polystyrene: ~42% in total pyrolyzate, ~13% in solid compounds in the pyrolyzate, ~5% in low-vapor-pressure liquid compounds in the pyrolyzate, and ~24% in high-vapor-pressure liquid compounds in the pyrolyzate.

4. Increase in $\dot{q}''_e$ in general reduces the amount of carbon converted to CO_2 and increases the amount of carbon converted to pyrolyzate, CO, and hydrocarbons, which is an indication of increased degradation of the fuel vapors.

Table 6 lists the data for the mean distribution of carbon in the combustion products of liquids and polymers. The chemical formula calculated from the elemental composition of the liquids and polymers are also listed in Table 6. The fuels have been arranged in decreasing order of f_{CO_2} values in Table 6. The carbon in the fuel converted to CO_2 decreases and the carbon converted to CO, hydrocarbons, and pyrolyzate increases from aliphatic to aromatic to highly chlorinated types of fuels.

In aliphatic-type fuels containing oxygen, the carbon in the fuel converted to CO_2 is higher and the carbon in the fuel converted to CO, hydrocarbons, and pyrolyzate is lower than the aliphatic-type fuels containing no oxygen. The tendency of the conversion of carbon in the fuel to various products appears to depend on the chemical structure of the fuels rather than on the ratios of C, H, N, O, and Cl atoms in the fuel. For example, from the ignition data in Table 1, the kinetic parameter $\phi \approx 0.28$. 0.14, and 0.12 for aliphatic, aromatic, and highly chlorinated types of fuels, respectively; the f_i values in Table 6 follow these ϕ values approximately. The higher the value of ϕ, the higher is the reactivity of fuel vapor for oxygen, i.e., higher f_{CO_2} values and lower f_{CO}, f_{HC}, and f_{pyr} values. The reactivity of the polymer vapors for oxygen, however, can be modified by changing the values of $\dot{q}''_e$ and m_{O_2} to increase the degradation of the fuel vapors in the preflame and flame zone. Increase in the degradation of fuel vapors is also helpful in increasing the flame radiation. Since the reactivity of aliphatic-type polymers, especially oxygenated polymers, is high, high values of $\dot{q}''_e$ and m_{O_2} are required to increase the degradation of fuel vapors.

TABLE 6

Mean Distribution of Carbon in the Combustion Products of Liquids and Polymers[a]

Liquid–Polymer	Chemical formula from elemental composition	Gases, $f_i \times 100$			Pyrolyzate, $f_i \times 100$		
		CO_2	CO	HC	Pyr[b]	Pyr-f[c]	$(Pyr)_{liquid\text{-}1}$[d]
Methanol	CH_4O	99.3	0.1	0	0.6	0	0.6
Polymethylmethacrylate	$CH_{1.6}O_{0.4}$	78.6	0.8	0.1	20.5	4.0	16.5
Polyoxymethylene	CH_2O	76.5	0.2	0.1	23.2	–	–
Acetone	$CH_2O_{0.33}$	76.2	0.1	0	23.7	–	–
Cellulose	$CH_{0.83}O_{0.33}$	71.6	0.5	0.1	27.8	–	–
Nylon 6/6	$CH_{1.83}O_{0.17}N_{0.17}$	71.5	2.6	2.1	23.8	21.8	2.0
Polyethylenes	CH_2	68.5	2.0	1.2	28.3	13.0	15.3
Heptane	$CH_{2.29}$	66.8	3.8	5.7	23.7	–	–
Polyurethane foam–$CaCO_3$ (I-A)	$CH_{1.91}O_{0.26}N_{0.06}$	65.6	0.6	0.1	33.7	30.0	3.7
Polypropylene	CH_2	64.4	1.8	2.3	31.5	–	–
Polyurethane foam (GM-21)	$CH_{1.80}O_{0.29}N_{0.05}$	60.8	0.7	0.4	38.1	31.6	6.5
Polyurethane foam (GM-25)	$CH_{1.75}O_{0.32}N_{0.07}$	59.4	2.3	0.7	37.6	31.0	6.6
Polyurethane foam (GM-23-FR)	$CH_{1.76}O_{0.35}N_{0.06}$	59.0	2.1	0.6	38.3	32.9	5.4
Polyethylene–25%Cl	$CH_{1.87}Cl_{0.13}$	57.1	3.8	2.4	36.7	39.3	0
Styrene	CH	54.0	3.0	2.2	40.8	39.7	1.1
Polystyrene	CH	51.9	2.9	2.8	42.4	19.0	23.4
Polyurethane foam (GM-27/FR)	$CH_{1.71}O_{0.30}N_{0.08}$	51.7	4.0	1.6	42.7	36.6	6.1
Polystyrene foam (GM-47)	$CH_{1.01}$	48.2	2.7	1.0	48.1	26.8	21.3
Benzene	CH	47.1	5.4	3.3	44.2	–	–
Styrene–Butadiene	$CH_{1.01}$	47.1	2.3	1.1	49.5	31.4	18.1
Aniline	$CH_{1.17}N_{0.17}$	46.3	2.7	1.2	49.8	–	–
Epoxy–FR–fiberglass	$CH_{1.32}N_{0.12}$	45.4	2.3	0.6	51.7	13.6	38.1
Polyurethane foam (GM-31-FR)	$CH_{1.17}O_{0.22}N_{0.10}$	40.7	5.1	1.9	52.3	51.2	1.1
Polyurethane foam (GM-29)	$CH_{1.15}O_{0.23}N_{0.10}$	40.1	5.4	1.5	53.0	32.6	20.4
Polyvinyl chloride	$CH_{1.5}Cl_{0.5}$	32.7	6.3	5.5	55.5	55.5	0
Polyethylene–36%Cl	$CH_{1.78}Cl_{0.22}$	31.6	3.9	2.7	61.8	63.6	0
Polyethylene–48%Cl	$CH_{1.65}Cl_{0.36}$	26.8	4.6	2.9	65.7	–	–

[a] Fires not fully ventilated.
[b] Total pyrolyzate calculated from Eq. (25).
[c] Pyr-f pyrolyzate fraction collected on the filter paper consisting of low-vapor pressure liquids and solid compounds.
[d] $(Pyr)_{liquid\text{-}1}$: high-vapor-pressure liquids in the pyrolyzate which cannot be collected on the filter paper calculated from Eqs. (25) and (26).

5.4. Conclusions

For evaluating the fire hazard of fuels in terms of mass generation rates of toxic compounds, $\dot{M}''_i$, and the f_i values are very useful.* In general the conversion of carbon in the vaporized fraction of the fuel to CO_2 is of the order of 72%, 51%, and 30% for the aliphatic, aromatic, and highly chlorinated types of polymers, respectively. The conversion of carbon in the vaporized fraction of the fuel to CO is of the order of 1%, 3%, and 5% for the aliphatic, aromatic, and highly chlorinated types of polymers. The conversion of carbon in the vaporized fraction of the fuel to gaseous hydrocarbons is of the order of 1%, 2%, and 4% for the aliphatic, aromatic, and highly chlorinated types of polymers. The conversion of carbon in the vaporized fraction of the fuel to pyrolyzate is of the order of 26%, 44%, and 61% for the aliphatic, aromatic, and highly chlorinated types of polymers, respectively. The majority of the carbon in the fuels is, thus, converted to CO_2 and pyrolyzate.

The conversion of carbon to CO_2, CO, hydrocarbons, and pyrolyzate depends on the reactivity of the fuel vapor for oxygen (i.e., chemical structure of the fuel) rather than on the ratio of C, H, N, O, and Cl atoms in the fuel. The reactivity of fuel vapor for oxygen, expressed as a kinetic parameter ϕ is $\approx$0.28, 0.14, and 0.12 for the aliphatic, aromatic, and highly chlorinated types of fuels, respectively. The reactivity of fuel vapor for oxygen can be modified by increasing $\dot{q}''_e$ and m_{O_2}.

6. Heat Release Rate

6.1. Concept

The theoretical stoichiometric heat release rate for complete combustion of the fuel is defined as

$$\dot{Q}''_{\text{th}} = \dot{m}''_b H_c \tag{27}$$

where $\dot{Q}''_{\text{th}}$ is the theoretical stoichiometric heat release rate for complete combustion of the fuel per unit sample surface area (kW/m^2), $\dot{m}''_b$ is the mass loss rate per unit sample surface area in combustion (g/m^2 sec), and H_c is the net heat of complete combustion of the fuel (kJ/g).

The combustion efficiency of fuel vapors can be defined as follows:

$$\chi_{\text{actual}} = \frac{Q''_{\text{actual}}}{\dot{Q}''_{\text{th}}} = \frac{\dot{Q}''_{\text{actual}}}{\dot{m}''_b H_c} \tag{28}$$

*$\dot{M}''_i = f_i k_i \cdot \dot{m}''_b$; f_i values are given in Table 5, k_i values can be calculated from chemical formulae in Table 6, and $\dot{m}''_b$ can be calculated from Eq. (13) and data in Table 3 or $\dot{m}''_b$ values can be measured directly.

where χ_{actual} is the combustion efficiency of the fuel vapors or the actual fraction of the theoretical stoichiometric heat release and $\dot{Q}''_{\text{actual}}$ the actual heat release rate per unit sample surface area (kW/m^2).

The actual heat release rate has two components, i.e., convective and radiative, and can be expressed as follows:

$$\dot{Q}''_{\text{actual}} = \dot{Q}''_{\text{conv}} + \dot{Q}''_{\text{rad}} \tag{29}$$

where $\dot{Q}''_{\text{conv}}$ is the convective heat release rate per unit sample surface area (kW/m^2) and $\dot{Q}''_{\text{rad}}$ is the radiative heat release rate per unit sample surface area (kW/m^2).

Dividing both sides of Eq. (29) by $\dot{Q}''_{\text{th}}$ or $\dot{m}''_b H_c$,

$$\chi_{\text{actual}} = \chi_{\text{conv}} + \chi_{\text{rad}} \tag{30}$$

where χ_{conv} is the convective fraction of the theoretical stoichiometric heat release rate and χ_{rad} the radiative fraction of the theoretical stoichiometric heat release rate.

$\dot{Q}''_{\text{conv}}$ or χ_{conv} can be measured in the experiments using Eq. (2) and $\dot{Q}''_{\text{actual}}$ or χ_{actual} can be calculated by the following method.

6.1.1. *Calculation of Actual Heat Release Rate*

The actual heat release rate can be calculated from the following two different sets of experimental data.

6.1.1.1. Actual Heat Release Rate from Production of CO and CO_2. The net heat of complete combustion of the fuel per unit mass of CO_2 produced can be expressed as H_c/k_{CO_2}, where k_{CO_2} is the theoretical stoichiometric mass of CO_2 produced per unit mass of the fuel converted completely to CO_2. The heat release rate for the production of CO_2 is thus equal to $(H_c/k_{CO_2})\dot{M}''_{CO_2}$ or $f_{CO_2}\dot{m}''_b H_c$, where $\dot{M}''_{CO_2}$ is the mass generation rate of CO_2 and f_{CO_2} is the fractional theoretical stoichiometric yield of CO_2 (or fraction of carbon in the fuel converted to CO_2). The net heat of combustion of the fuel converted to CO per unit mass of CO produced is equal to $H_c/k_{CO} - H_{CO}$, where k_{CO} is the theoretical stoichiometric mass of CO produced per unit mass of the fuel converted completed to CO, and H_{CO} is the heat of combustion of CO. The heat release rate for the production of CO is thus equal to $(H_c/k_{CO} - H_{CO})\dot{M}''_{CO}$ or $f_{CO}\dot{m}''_b H_c - H_{CO}k_{CO}f_{CO}\dot{m}''_b$, where $\dot{M}''_{CO}$ is the mass generation rate of CO and f_{CO} is the fractional theoretical stoichiometric yield of CO (or fraction of carbon in the fuel converted to CO).

If we assume that in the combustion of fuels heat is released in the production of CO and CO_2, then the actual heat release rate can be expressed as

$$\dot{Q}''_{\text{actual}} \approx \dot{m}''_b[H_c(f_{CO} + f_{CO_2}) - H_{CO}f_{CO}k_{CO}] \tag{31}$$

Values of f_{CO} and f_{CO_2} are given in Table 5 and $\dot{m}''_{g}$ and H_c are measured in the experiments; thus $\dot{Q}''_{actual}$ can be calculated. Table 7 lists $\dot{Q}''_{actual}$ values calculated from Eq. (31).

6.1.1.2. Actual Heat Release Rate from O_2 Depletion. The net heat of complete combustion of the fuel per unit mass of O_2 consumed is equal to H_c/k_{O_2}, where k_{O_2} is the theoretical stoichiometric mass of O_2 consumed per

TABLE 7
Comparison of Actual Heat Release Rate Values Calculated from Two Different Sets of Experimental Data

Polymer–liquid	$\dot{Q}''_{actual}$ (kW/m²) From CO + CO$_2^a$ [Eq. (31)]	From O$_2^b$ [Eq. (32)]	$\dot{Q}''_{actual}$ from Eq. (31) / $\dot{Q}''_{actual}$ from Eq. (32)
Methanol (liquid)	325	323	1.006
	324	335	0.967
	443	438	1.011
	447	448	0.998
	532	546	0.974
	660	663	0.995
	704	741	0.950
	1106	1081	1.023
	1120	1123	0.997
	1144	1112	1.029
Acetone (liquid)	261	263	0.992
Heptane (liquid)	834	839	0.994
Benzene (liquid)	1215	1289	0.943
Sytrene (liquid)	2421	2444	0.991
Aniline (liquid)	449	357	1.257
	842	789	1.067
Polyethylene rigid foams			
1	118	163	0.724
	134	126	1.063
	252	277	0.910
	263	273	0.963
Polyethylene rigid foams			
1	347	349	0.994
Polyethylene rigid foams			
2	217	249	0.871
	342	342	1.000
	447	493	0.907
	475	501	0.948
3	251	236	1.064
	447	493	0.907
	480	491	0.978
	589	589	1.000

TABLE 7 (Continued)

Polymer–liquid	$\dot{Q}''_{actual}$ (kW/m^2) From CO + CO$_2$[a] [Eq. (31)]	From O$_2$[b] [Eq. (32)]	$\dot{Q}''_{actual}$ from Eq. (31) / $\dot{Q}''_{actual}$ from Eq. (32)
4	304	299	1.017
	474	537	0.883
	567	564	1.005
	726	736	0.986
Polyurethane flexible foams			
GM-21	353	341	1.035
GM-23	386	351	1.100
Granular polymers			
Polyoxymethylene	265	272	0.974
Polystyrene	584	652	0.896
	642	733	0.876
Polymethylmethacrylate	721	760	0.949
Polypropylene	1225	1205	1.017
			Mean 0.982
			Standard deviation 0.0804

[a]Equation (31), $\dot{Q}''_{actual} = \dot{m}''_b[H_c(f_{CO} + f_{CO_2}) - H_{CO}k_{CO}f_{CO}]$.
[b]Equation (32), $\dot{Q}''_{actual} = H_c f_{O_2}\dot{m}''_b$.

unit mass of fuel for complete combustion. The heat release rate for the consumption of oxygen then is given by $(H_c/k_{O_2})\dot{M}''_{O_2}$ or $H_c f_{O_2}\dot{m}''_b$, where $\dot{M}''_{O_2}$ is the mass depletion rate of oxygen and f_{O_2} is the fractional theoretical stoichiometric consumption of O_2.

If we assume that in the combustion of fuels, heat is released in the consumption of O_2, then the actual heat release rate can be expressed as follows:

$$\dot{Q}''_{actual} \approx \frac{H_c}{k_{O_2}}\dot{M}''_{O_2} \approx H_c f_{O_2}\dot{m}''_b \tag{32}$$

In the experiments H_c, k_{O_2}, $\dot{M}''_{O_2}$, and f_{O_2} are measured and $\dot{Q}''_{actual}$ thus can be calculated. Table 7 lists the $\dot{Q}''_{actual}$ values calculated from Eq. (32).

A comparison of $\dot{Q}''_{actual}$ values in Table 7, calculated from two different sets of experimental data, shows excellent agreement. The mean value of $\dot{Q}''_{actual}$ calculated from Eq. (31) is 98% of the mean value calculated from Eq. (32). Thus, for the calculation of $\dot{Q}''_{actual}$ values either of the equations can be used. In our experiments, we use an average value. We can also write

$$H_c f_{O_2}\dot{m}''_b = \dot{m}''_b[H_c(f_{CO} + f_{CO_2}) - H_{CO}f_{CO}k_{CO}] \tag{33}$$

or

$$f_{O_2} = f_{CO} + f_{CO_2} - \frac{H_{CO}}{H_c} f_{CO} k_{CO} \tag{34}$$

or

$$\frac{f_{O_2}}{f_{CO} + f_{CO_2}} = 1 - \frac{H_{CO} f_{CO} k_{CO}}{H_c(f_{CO} + f_{CO_2})} \tag{35}$$

In Eq. (35), f_{CO} and $H_{CO} f_{CO} k_{CO} / H_c(f_{CO} + f_{CO_2})$ depend very much on the generic nature of the polymers and the magnitudes of $\dot{q}''_e$ and $\dot{m}_{O_2}$. For example, for aliphatic-type polymers, especially the oxygenated polymers, $f_{CO} \ll f_{CO_2}$ and $H_{CO} f_{CO} k_{CO} / H_c(f_{CO} + f_{CO_2}) \ll 1.0$ and thus $f_{O_2} \approx f_{CO_2}$. As $\dot{q}''_e$ and m_{O_2} are increased, however, the values of f_{CO} and $H_{CO} f_{CO} k_{CO} / H_c(f_{CO} + f_{CO_2})$ increase.

The available data for H_c, k_{O_2}, and k_{CO_2} show that H_c/k_{O_2} and H_c/k_{CO_2} are approximately constants for gases, liquids, and polymers irrespective of their generic classes, as shown in Table 8. Except for silicones, it appears that a

TABLE 8
H_c/k_{O_2} and H_c/k_{CO_2} Values for Common Liquids and Polymers

Fuel	H_c/k_{O_2} (kJ/g)	H_c/k_{CO_2} (kJ/g)
Granular polymers		
Polymethylmethacrylate	13.2	11.5
Polystyrene	12.7	11.6
Polypropylene	12.6	13.9
Polyethylene (PE)	12.7	13.9
PE-25%Cl	12.7	13.4
PE-36%Cl	12.8	12.9
PE-36%Cl	12.8	12.9
PE-48%Cl	12.8	12.3
Polyvinyl chloride	12.8	11.7
Nylon 6/6	11.8	13.2
Styrene-butadiene	13.3	12.2
Polyoxymethylene	14.5	10.6
Mean	12.9	12.5
Standard deviation	0.65	1.08
Silicones[a]		
Linear		
MM	16.2	20.6
MDM	15.5	20.4
MD_2M	13.7	18.5
MD_3M	15.4	19.5
Cyclics		
D_3	14.2	20.6

Fuel	H_c/k_{O_2} (kJ/g)	H_c/k_{CO_2} (kJ/g)
D_4	14.7	21.4
D_5	14.6	21.3
D_6	14.6	21.3
Mean	14.9	20.5
Standard deviation	0.80	1.01
Foamed polymers		
Polyurethanes		
GM-21	11.6	11.5
GM-23-FR	13.0	12.5
GM-25	11.3	11.1
GM-27-FR	10.4	10.3
GM-29	11.5	10.7
GM-31	11.0	10.2
GM-37	11.8	11.2
Polyisocyanurates		
GM-41	11.3	10.4
GM-43	9.8	8.9
Polystyrenes		
GM-47	12.3	11.3
GM-49	12.4	11.4
Polystyrenes		
GM-51	11.6	10.8
GM-53	12.3	11.3
Phenolic Foam		
GM-57	(16.8)	14.1
Polyethylenes		
1	12.1	13.2
2	11.8	13.1
3	12.0	13.0
4	12.0	13.0
Mean	11.7	11.6
Standard deviation	0.77	1.33
Liquids		
Pentane	12.5	14.5
Hexane	12.7	14.6
Heptane	12.7	14.5
Benzene	13.0	11.9
Toluene	12.9	12.1
Benzaldehyde	13.2	11.0
Aniline	11.6	12.3
Styrene	13.1	12.0
Methyl alcohol	13.3	14.5
Ethyl alcohol	12.8	14.0
Propyl alcohol	12.7	13.9
Acetone	12.9	12.5
Acetaldehyde	13.6	12.4
Formaldehyde	14.5	10.6
Mean	13.0	12.9
Standard deviation	0.64	1.38

[a] $M = (CH_3)_3SiO_{0.5}$, $D = (CH_3)_2SiO$.

mean value of 12.5 kJ/g can be used in Eq. (32) for H_c/k_{O_2}, i.e.,

$$\dot{Q}''_{\text{actual}} \approx 12.5\dot{M}''_{O_2} \qquad (\text{kW/m}^2) \tag{36}$$

The usefulness of Eq. (36) is that actual heat release rates can be calculated by measuring $\dot{M}''_{O_2}$ alone independent of shape, size, orientation, generic nature of the polymers, mixtures of polymers, in the presence and absence of the extinguishing agents, etc.

The limitation of Eq. (36), however, is the accuracy with which $\dot{M}''_{O_2}$ can be measured, especially in fires where products are diluted with large volumes of air.

The data in Table 8 also indicate that, for granular and foamed polymers and liquids, $H_c/k_{CO_2} \approx 12.3$. For conditions where $f_{CO} \ll f_{CO_2}$ and $H_{CO}f_{CO}k_{CO}/H_c(f_{CO}+f_{CO_2}) \ll 1.00$, $\dot{Q}''_{\text{actual}}$ can be calculated from the following equation:

$$\dot{Q}''_{\text{actual}} \approx 12.3\ \dot{M}''_{O_2} \qquad (\text{kW/m}^2) \tag{37}$$

Equation (37) is as useful at Eq. (36); however, it is limited because of larger variations in the H_c/k_{CO_2} values compared to H_c/k_{O_2}, and an error of about 6% in $\dot{Q}''_{\text{actual}}$ is introduced (based on our data) if f_{CO} and $H_{CO}f_{CO}k_{CO}/H_c(f_{CO}+f_{CO_2})$ are not negligible. On the other hand $\dot{M}''_{CO_2}$ can be measured much more accurately than $\dot{M}''_{O_2}$, irrespective of the dilution of products by fresh air. In our studies we are planning to use both Eqs. (36) and (37) for larger-scale fires.

Since $\dot{Q}''_{\text{actual}}$ and $\dot{Q}''_{\text{conv}}$ values are known, $\dot{Q}''_{\text{rad}}$ values can be calculated from Eq. (29).

6.2. *Measurements and Calculations*

$\dot{Q}''_{\text{conv}}$: The convective heat release rate per unit sample surface area is obtained from Eq. (2), where $\dot{M}''_i$, T_d, and T_a are measured and $C_p(T)$ value is used from the literature.

The heat losses in our apparatus are negligibly small and furthermore they are eliminated because a difference is taken in Eq. (2). $\dot{Q}''_{\text{conv}}$, however, is expected to be slightly higher than the true value because some flame radiation is captured in the apparatus and is included in $\dot{Q}''_{\text{conv}}$; χ_{conv} is then calculated from $\dot{Q}''_{\text{conv}}/\dot{m}''_{g}H_c$.

$\dot{Q}''_{\text{actual}}$: The actual heat release rate per unit sample surface area is obtained from Eqs. (31) and (32) and averaged. All the quantities needed in Eqs. (31) and (32) are measured in the experiments. χ_{actual} is then calculated from $\dot{Q}''_{\text{actual}}/\dot{m}''_{g}H_c$.

$\dot{Q}''_{\text{rad}}$: The radiative heat release rate per unit sample surface area is obtained from Eq. (29). The value of $\dot{Q}''_{\text{rad}}$ obtained in this fashion is expected to be slightly less than the true value, because some of the radiation captured in the apparatus is included in $\dot{Q}''_{\text{conv}}$; χ_{rad} is then calculated from $\dot{Q}''_{\text{rad}}/\dot{m}''_{g}H_c$.

6.3. Data for the Heat Release Rate Fractions

The data for the heat release rate fractions for liquids and polymers are listed in Table 9.

The χ_{actual} values follow the fraction of carbon in the fuel converted to CO_2 (Table 5) very closely. χ_{actual} decreases from aliphatic to aromatic to highly chlorinated fuels. The combustion efficiencies of aliphatic, aromatic, and highly chlorinated fuels are of the order of 71%, 57%, and 33%,

TABLE 9
Fraction of Theoretical Stoichiometric Heat Release Rate for the Combustion of Liquids and Polymers[a]

Polymer–liquid	$\dot{q}''_e$ (kW/m^2)	m_{O_2}	χ_{actual}	χ_{conv}	χ_{rad}
Aliphatic-type fuels					
Liquids					
Methanol	0	0.233	0.993	0.853	0.141
Acetone	0	0.233	0.762	0.566	0.196
Heptane	0	0.233	0.690	0.374	0.316
Polymers					
Cellulose	52.4	0.233	0.716	0.351	0.365
Polyoxymethylene	0	0.233	0.755	0.607	0.148
	38.9	0.233	0.777	0.560	0.217
	52.4	0.233	0.764	0.542	0.222
Polymethylmethacrylate	0	0.233	0.867	0.622	0.245
	39.7	0.233	0.710	0.340	0.360
	52.4	0.233	0.710	0.410	0.300
Polypropylene	0	0.233	0.752	0.548	0.204
	39.7	0.233	0.593	0.233	0.360
	52.4	0.233	0.679	0.267	0.413
Nylon 6/6	52.4	0.233	0.728	0.283	0.445
Polyethylenes (PE)					
PE	0	0.233	0.722	0.536	0.190
	40.2	0.233	0.739	0.257	0.482
PE foams					
1	38.9	0.233	0.700	0.570	0.130
	52.0	0.233	0.685	0.485	0.200
2	26.5	0.233	0.790	0.640	0.150
	38.9	0.233	0.720	0.510	0.210
	52.0	0.233	0.700	0.480	0.220
3	26.5	0.233	0.730	0.520	0.210
	38.9	0.233	0.700	0.480	0.220
	52.0	0.233	0.640	0.440	0.200
4	26.5	0.233	0.740	0.590	0.150
	38.9	0.233	0.700	0.460	0.240
	52.0	0.233	0.559	0.460	0.099
PE–25%Cl	35.1	0.233	0.620	0.159	0.461
	38.9	0.233	0.571	0.196	0.375

TABLE 9 (Continued)

Polymer–liquid	$\dot{q}_e''$ (kW/m^2)	m_{O_2}	χ_{actual}	χ_{conv}	χ_{rad}
PE–36%Cl	26.8	0.233	0.336	0.176	0.160
PE–48%Cl	20.1	0.233	0.291	0.221	0.070
Polyvinyl chloride	52.4	0.233	0.357	0.148	0.209
Aromatic-type fuels					
Liquids					
Aniline	0	0.233	0.640	0.410	0.230
	20.0	0.233	0.470	0.200	0.270
Styrene	0	0.233	0.550	0.180	0.370
Benzene	0	0.233	0.529	0.170	0.359
Polymers					
Polystyrene	0	0.233	0.607	0.385	0.222
	32.5	0.233	0.392	0.090	0.302
	38.9	0.233	0.413	0.130	0.283
	39.7	0.233	0.464	0.130	0.334
Epoxy–FR–fiberglass	38.9	0.233	0.510	0.131	0.379
	47.0	0.233	0.463	0.137	0.326
	52.4	0.233	0.518	0.148	0.370
		0.279	0.483	0.067	0.416
		0.334	0.468	0.072	0.396
Polyurethane foams					
I-A–$CaCO_3$	0	0.233	0.683	0.354	0.329
	10.0	0.233	0.697	0.319	0.378
	20.1	0.233	0.639	0.327	0.312
	26.4	0.233	0.620	0.328	0.292
	38.9	0.233	0.674	0.263	0.411
	45.2	0.233	0.652	0.193	0.459
	52.3	0.233	0.641	0.227	0.414
	59.8	0.233	0.666	0.220	0.446
GM-21	45.2	0.160	0.616	0.312	0.304
		0.180	0.645	0.288	0.357
		0.213	0.638	0.251	0.387
		0.233	0.622	0.221	0.400
GM-23–FR	45.2	0.167	0.615	0.194	0.421
		0.233	0.612	0.167	0.445
		0.273	0.631	0.151	0.480
GM-25	26.4	0.207	0.617	0.310	0.307
		0.214	0.656	0.315	0.341
		0.233	0.638	0.249	0.389
		0.301	0.609	0.171	0.438
	31.8	0.233	0.580	0.178	0.402
	38.9	0.233	0.579	0.201	0.378
	52.4	0.233	0.585	0.181	0.404
	59.8	0.233	0.597	0.219	0.378
GM-27–FR	31.8	0.233	0.534	0.212	0.322
	38.9	0.233	0.520	0.215	0.305
	52.4	0.233	0.518	0.185	0.333
GM-29	52.0	0.233	0.455	0.148	0.307
GM-31–FR	52.0	0.233	0.458	0.151	0.307

[a]Fires not fully ventilated.

respectively. The combustion efficiency follows the reactivity of fuel vapors for oxygen expressed as a kinetic parameter ϕ (in Table 1), i.e., $\phi \approx 0.28, 0.14$, and 0.12 for aliphatic, aromatic, and highly chlorinated fuels, respectively. Since the reactivity of the fuel vapor for oxygen can be modified by increasing $\dot{q}''_e$ and m_{O_2} values, χ_{actual} can also be modified.

The χ_{rad} value for fuels burning with nonluminous or slightly luminous flames is equal to about 0.18 irrespective of the generic nature of the fuels. For fuels burning with luminous flames, $\chi_{rad} \approx 0.37$. For the combustion of polymers in 0.093-m^2 horizontal pools, measured values of χ_{rad} have been reported to be equal to 0.15, 0.34, 0.38, 0.35, and 0.32 for polyoxymethylene, polymethylmethacrylate, polypropylene, polystyrene, and flexible polyurethane foam–$CaCO_3$(I-A), respectively.[11,12] These reported χ_{rad} values show agreement with the χ_{rad} values in Table 9 for these fuels. Thus, our data indicate that χ_{rad} values are independent of the generic nature of the fuels and can be derived reasonably accurately and quite simply from our technique.

χ_{conv} decreases as $\dot{q}''_e$ and m_{O_2} are increased and the magnitude of decrease depends on the generic nature of the fuels. Table 10 lists $\chi_{conv}/\chi_{actual}$ values for the combustion of polymers at $\dot{q}''_e > 38.9$ kW/m^2 in normal air; data for benzene and styrene at $\dot{q}''_e = 0$ are also included in the table. For oxygenated aliphatic-type polymers the ratio is high, and for aromatic-type polymers the ratio is low. It appears that, for producing highly radiative flames, m_{O_2} is required to be increased substantially above the value for normal air (0.233)

TABLE 10
$\chi_{conv}/\chi_{actual}$ Values for the Combustion of Polymers at $\dot{q}''_e \geqslant 38.9$ kW/m^2 in Normal Air

Polymer	$\chi_{conv}/\chi_{actual}$
Polyoxymethylene	0.715
Polymethylmethacrylate	0.528
Polypropylene	0.395
Nylon 6/6	0.389
Polyurethane foam (GM-27-FR)	0.389
Polyurethane foam (GM-21)	0.355
Polyethylene	0.348
Polyurethane foam (GM-25)	0.344
Polyurethane foam (I-A–$CaCO_3$)	0.343
Polyurethane foam (GM-31-FR)	0.330
Styrene[a] ($\dot{q}''_e = 0$)	0.327
Polyurethane foam (GM-29)	0.325
Benzene[a] ($\dot{q}''_e = 0$)	0.321
Polystyrene	0.275
Polyurethane foam (GM-23-FR)	0.273
Epoxy-FR-glass fibers	0.266

[a]Liquids.

TABLE 11
Net Heat of Combustion of Fuels

Polymer–Liquid	$\dot{q}_e''$ (kW/m^2)	m_{O_2}	Net heat of combustion (kJ/g) Complete combustion	Actual[a]
Aliphatic-type fuels				
Liquids				
Methanol	0	0.233	19.96	19.82
Acetone	0	0.233	28.49	21.71
Heptane	0	0.233	44.56	30.75
Polymers				
Cellulose	52.4	0.233	16.48	11.80
Polyoxymethylene	0	0.233	15.48	11.69
	38.9	0.233	15.48	12.03
	52.4	0.233	15.48	11.83
Polymethylmethacrylate	0	0.233	25.20	21.85
	39.7	0.233	25.20	17.89
	52.4	0.233	25.20	17.89
Polypropylene	0	0.233	43.38	32.62
	39.7	0.233	43.38	25.72
	52.4	0.233	43.38	29.46
Nylon 6/6	52.4	0.233	30.79	22.42
Polyethylenes (PE)				
PE	0	0.233	43.47	31.39
	40.2	0.233	43.47	32.12
PE foams				
1	38.9	0.233	41.21	28.85
	52.0	0.233	41.21	28.23
2	26.5	0.233	40.84	32.26
	38.9	0.233	40.84	29.40
	52.0	0.233	40.84	28.59
3	26.5	0.233	40.84	29.81
	38.9	0.233	40.84	28.59
	52.0	0.233	40.84	26.14
4	26.5	0.233	40.84	30.22
	38.9	0.233	40.84	28.59
	52.0	0.233	40.84	22.83
PE–25%Cl	35.1	0.233	31.59	19.59
	38.9	0.233	31.59	18.04
PE–36%Cl	26.8	0.233	26.28	8.83
PE–48%Cl	20.1	0.233	20.63	6.00
Polyvinyl chloride	52.4	0.233	16.44	5.89
Aromatic-type fuels				
Liquids				
Aniline	0	0.233	34.86	22.30
	20.0	0.233	34.86	16.38
Styrene	0	0.233	40.51	22.28
Benzene	0	0.233	40.14	21.23

TABLE 11 (Continued)

Polymer–Liquid	$\dot{q}_e''$ (kW/m²)	m_{O_2}	Net heat of combustion (kJ/g)	
			Complete combustion	Actual[a]
Polymers				
Polystyrene	0	0.233	39.17	23.77
	32.5	0.233	39.17	15.35
	38.9	0.233	39.17	16.18
	39.7	0.233	39.17	18.17
Epoxy–FR–Fiberglass	38.9	0.233	29.83	15.21
	47.0	0.233	29.83	13.81
	52.4	0.233	29.83	15.45
		0.279	29.83	14.41
		0.334	29.83	13.96
Polyurethane foams				
I-A–$CaCO_3$	0	0.233	17.11	11.69
	10.0	0.233	17.11	11.93
	20.1	0.233	17.11	10.93
	26.4	0.233	17.11	10.61
	38.9	0.233	17.11	11.53
	45.2	0.233	17.11	11.16
	52.3	0.233	17.11	10.97
	59.8	0.233	17.11	11.40
GM-21	45.2	0.160	26.15	16.11
		0.180	26.15	16.87
		0.213	26.15	16.68
		0.233	26.15	16.27
GM-23-FR	45.2	0.167	27.15	16.70
		0.233	27.15	16.62
		0.273	27.15	17.13
GM-25	26.4	0.207	24.64	15.20
		0.214	24.64	16.16
		0.233	24.64	15.72
		0.301	24.64	15.01
	31.8	0.233	24.64	14.29
	38.9	0.233	24.64	14.27
	52.4	0.233	24.64	14.41
	59.8	0.233	24.64	14.71
GM-27-FR	31.8	0.233	23.18	12.38
	38.9	0.233	23.18	12.05
	52.4	0.233	23.18	12.01
GM-29	52.0	0.233	26.02	11.84
GM-31-FR	52.0	0.233	25.02	11.46

[a]Fires not fully ventilated.

for aliphatic-type polymers, especially oxygenated polymers, as well as for the polyurethane foams such that $\chi_{\text{conv}}/\chi_{\text{actual}} \approx 0.26$. Currently, we are examining various fuels over a wide range of m_{O_2} values to achieve the asymptotic value of $\dot{q}''_{\text{fr}}/L_g$ (see Table 3), where $\chi_{\text{conv}}/\chi_{\text{actual}}$ will also be examined. Note that $\chi_{\text{rad}} = 1 - (\chi_{\text{conv}}/\chi_{\text{actual}})$.

Since χ_{actual} is the combustion efficiency of the fuel, the net heat of complete combustion times χ_{actual} is the actual heat of combustion of the fuel. Table 11 lists the data for the net heat of complete combustion and actual heat of combustion of fuels. If $\dot{m}''_g$ values can be measured or predicted, calculation of actual heat release rate is straightforward.

6.4. Conclusions

For evaluating the fire hazard of polymers in terms of heat release rates, χ_i values are useful. The combustion efficiencies of aliphatic, aromatic, and highly chlorinated types of fuels are of the order of 71%, 57%, and 33%, respectively. The combustion efficiency follows the reactivity of fuel vapor for oxygen. χ_{rad} values for fuels burning with nonluminous or slightly luminous flames are of the order 0.18 and for fuels burning with luminous flame, of the order 0.37, χ_{conv} values depend on the generic nature of fuels. For highly radiative flames, $\chi_{\text{conv}}/\chi_{\text{actual}}$ is of the order of 0.26. Therefore, it appears that, if a polymer is burned under conditions where the combustion is highly radiative, about 74% of the actual heat will be released as radiation and $\dot{q}''_{\text{fr}}/L_g$ (see Table 3) would be expected to reach the asymptotic value.

7. Optical Transmission Through the Mixture of Pyrolysis–Combustion Products and Air

7.1. Concept

From Bouger's law[13]

$$\ln \frac{I_0}{I} = \frac{3}{2} \frac{\bar{K}(C_{\text{pyr}})l}{\rho D_{32}} \tag{38}$$

where, I_0 is the optical transmission through air (mV), I the optical transmission through the products (mV), l the optical path length (m), $\bar{K}$ the dimensionless average extinction coefficient, C_{pyr} the mass concentration of the pyrolyzate responsible for absorption of light (g/m^3), ρ the density of individual particles (g/m^3), and D_{32} the volume-to-surface mean diameter (m). Equation (38) can also be expressed as follows:

$$\sigma_{\text{pyr}} = \frac{1}{C_{\text{pyr}}\, l} \ln \frac{I_0}{I} \tag{39}$$

$$\sigma_{\text{pyr}} = \frac{3}{2} \frac{K}{\rho D_{32}} \tag{40}$$

where, σ_{pyr} is the mass absorbency index of the pyrolyzate (g/m^2).

C_{pyr} in Eq. (39) can be expressed as follows:

$$C_{\text{pyr}} = \frac{\dot{M}''_{\text{pyr}}\, a}{\dot{V}_T} \tag{41}$$

where, $\dot{M}''_{\text{pyr}}$ is the mass generation rate of the pyrolyzate per unit sample surface area (g/m^2 sec), a is the sample surface area (m^2), and $\dot{V}_T$ is the total volumetric flow rate of the mixture of products and air (m^3/sec). $\dot{M}''_{\text{pyr}}$ in Eq. (41) can be expressed as follows:

$$\dot{M}''_{\text{pyr}} = f_{\text{pyr}}\, k_{\text{pyr}}\, \dot{m}''_b \tag{42}$$

where k_{pyr} is the theoretical stoichiometric mass of the pyrolyzate produced per unit mass of the fuel converted completely to the pyrolyzate f_{pyr} the fractional theoretical stoichiometric yield of the pyrolyzate (or the fraction of carbon in the fuel converted to pyrolyzate), and $\dot{m}''_b$ is the mass loss rate of the fuel per unit sample surface area (g/m^2 sec).

From Eqs. (41) and (42),

$$C_{\text{pyr}} = \frac{f_{\text{pyr}}\, k_{\text{pyr}}\, \dot{m}_b}{\dot{V}_T} \tag{43}$$

where $\dot{m}_b$ is the mass loss rate of the fuel (g/sec).

From Eqs. (39) and (43),

$$\sigma_{\text{pyr}}\, f_{\text{pyr}}\, k_{\text{pyr}} = \frac{\dot{V}_T}{\dot{m}_b\, l} \ln \frac{I_0}{I} \tag{44}$$

From Eqs. (26) and (44)

$$(\sigma_{\text{pyr}} f_{\text{pyr}} k_{\text{pyr}})_{\text{liquid-1}} + (\sigma_{\text{pyr}} f_{\text{pyr}} k_{\text{pyr}})_{\text{liquid-2}} + (\sigma_{\text{pyr}} f_{\text{pyr}} k_{\text{pyr}})_{\text{solid}} = \frac{\dot{V}_T}{\dot{m}_b\, l} \ln \frac{I_0}{I} \tag{45}$$

where liquid-1 represents the high-vapor-pressure liquid compounds in the pyrolyzate, liquid-2 the low-vapor-pressure liquid compounds in the pyrolyzate, and solid means solid compounds in the pyrolyzate.

By using the filter paper weighing technique, $(f_{\text{pyr}} k_{\text{pyr}})_{\text{liquid-2}}$ and $(f_{\text{pyr}} k_{\text{pyr}})_{\text{solid}}$ can be calculated; however, $(f_{\text{pyr}} k_{\text{pyr}})_{\text{liquid-1}}$ is unknown. Furthermore, variation of temperature can influence the optical transmission significantly if pyrolyzate contains significant amounts of liquid-1. Because of these complications, only Eq. (44) can be used where $\sigma_{pyr} f_{pyr} k_{pyr}$ can be defined as a "modified mass absorbency index" of the pyrolyzate, expressed as ξ_{pyr}. In Eq. (44), $\dot{m}_b / \dot{V}_T$ is the mass concentration of the fuel vapors expressed as C_{fuel} (g/m^3), and thus

$$\xi_{pyr} = \frac{1}{C_{fuel}\, l} \ln \frac{I_0}{I} \tag{46}$$

For proper comparison of ξ_{pyr} values of pyrolyzate generated from various polymers, experiments should be performed such that the temperature of the mixture of products and air is the same in all the experiments for different polymers (possibly close to ambient temperature).

The higher the value of ξ_{pyr}, the lower will be the optical transmission (or visibility). In addition, if the physical properties of the pyrolyzate do not change, ξ_{pyr} will be independent of $\dot{V}_T$, $\dot{m}_b$, or l.

7.2. *Measurements and Calculations*

The optical density per unit path length is calculated from Eq. (5), and $\dot{V}_T$ and $\dot{m}_b$ are measured in the experiments. Although products are diluted 20 times their volume by fresh air, unfortunately in our experiments the temperature of the mixture of products and air is significantly higher than ambient and varies significantly between pyrolysis and combustion experiments, for different $\dot{q}''_e$ and m_{O_2} values and for different polymers. Thus, ξ_{pyr} values calculated from Eq. (46) may not show a true comparison of pyrolyzate generated from different polymers, especially for those where significant amounts of high-vapor-pressure liquids are generated (see Table 5).

7.3. *Data for Modified Mass Absorbency Index of "Pyrolyzate"*

The data for C_{fuel}, $(1/l) \ln (I_0/I)$ and ξ_{pyr} are listed in Table 12 for various fuels.

In general ξ_{pyr} is smaller for aliphatic-type fuels and polyurethane foams compared to aromatic-type fuels. The chlorinated-type fuels have highest ξ_{pyr} values. ξ_{pyr} values are higher under pyrolysis conditions than under the combustion conditions. The higher the value of ξ_{pyr}, the lower is the optical transmission (or visibility) through the products.

The average values of ξ_{pyr} are listed in Table 13. A comparison of ξ_{pyr} values suggests that the pyrolyzate from oxygenated aliphatic-type polymers have lowest ξ_{pyr} values and pyrolyzate from aromatic-type polymers have highest ξ_{pyr} values. ξ_{pyr} values for the pyrolyzate from chlorinated polyeth-

TABLE 12
Mass Concentration of Fuel Vapor, Optical Density, and Modified Mass Absorbency Index of Pyrolyzate in the Pyrolysis–Combustion of Fuels

Polymer–Liquid	$\dot{q}_e''$ (kW/m^2)	m_{O_2}	Mode	c_{fuel} (g/m^3)	$\frac{1}{l}\ln I_o/I$ (m^{-1})	ξ_{pyr} (m^2/g)
Aliphatic-type fuels						
Cellulose	52.4	0.233	Comb	4.39	0.06	0.01
Polyoxymethylene	0	0.233	Comb	1.26	0	0
	38.9	0.233	Comb	3.03	0	0
	52.4	0.233	Comb	4.10	0.04	0.01
Polymethylmethacrylate	0	0.233	Comb	1.73	0.19	0.11
	39.7	0.233	Comb	5.33	1.38	0.26
	52.4	0.233	Comb	7.89	1.26	0.16
Polypropylene	0	0.233	Comb	1.28	0.56	0.44
	39.7	0.233	Comb	4.48	2.72	0.61
	52.4	0.233	Comb	7.92	3.04	0.38
Nylon 6/6	52.4	0.233	Comb	6.42	2.62	0.41
Polyethylenes (PE)						
PE	0	0.233	Comb	1.20	0.35	0.29
	21.0	0.233	Comb	4.10	1.76	0.43
	40.2	0.233	Pyr	0.49	0.25	0.51
	40.2	0.233	Comb	6.34	2.09	0.33
PE foams						
1	26.5	0.233	Comb	1.51	0.54	0.36
	38.9	0.233	Comb	1.73	1.40	0.81
	52.0	0.233	Comb	2.18	1.75	0.80
2	26.5	0.233	Comb	1.36	1.04	0.76
	38.9	0.233	Comb	2.09	1.66	0.79
	52.0	0.233	Comb	2.97	2.25	0.76
3	26.5	0.233	Comb	1.43	1.12	0.78
	38.9	0.233	Comb	3.09	2.42	0.78
	52.0	0.233	Comb	3.79	2.93	0.77
4	26.5	0.233	Comb	1.77	1.38	0.78
	38.9	0.233	Comb	3.25	2.42	0.74
	52.0	0.233	Comb	4.58	3.03	0.66
PE–25%Cl	35.1	0.233	Pyr	1.22	0.33	0.27
	35.1	0.233	Comb	3.85	4.35	1.13
	38.9	0.233	Comb	4.54	5.35	1.18
PE–36%Cl	26.4	0.233	Pyr	1.36	0.79	0.58
	26.8	0.233	Comb	3.55	4.84	1.36
PE–48%Cl	20.1	0.233	Comb	2.96	4.03	1.36
	26.4	0.233	Pyr	1.35	0.79	0.59
Polyvinyl chloride	52.4	0.233	Comb	5.45	5.53	1.01
Aromatic-type fuels						
Aniline (liquid)	0	0.233	Comb	3.19	2.42	0.76
	20.0	0.233	Comb	7.97	4.78	0.60
Styrene (liquid)	0	0.233	Comb	3.92	3.61	0.92

TABLE 12 (Continued)

Polymer–Liquid	$\dot{q}_e''$ (kW/m²)	m_{O_2}	Mode	C_{fuel} (g/m³)	$\frac{1}{l}\ln I_o/I$ (m⁻¹)	ξ_{pyr} (m²/g)
Benzene (liquid)	0	0.233	Comb	8.36	7.50	0.90
Styrene–butadiene	0	0.233	Comb	1.35	1.96	1.45
	45.2	0.233	Comb	3.98	4.09	1.03
Polystyrene	0	0.233	Comb	1.51	1.37	0.91
	32.5	0.233	Comb	6.32	6.79	1.07
	38.9	0.233	Comb	6.52	7.61	1.17
Epoxy–FR–fiberglass	38.9	0.233	Comb	5.35	1.51	0.28
	47.0	0.233	Comb	5.98	2.07	0.35
	52.4	0.233	Comb	5.91	2.19	0.37
		0.279	Comb	9.32	2.07	0.22
		0.334	Comb	9.16	2.28	0.25
	59.7	0.233	Comb	5.44	1.65	0.30
Polystyrene foam (GM-47)	0	0.233	Comb	1.93	1.82	0.94
Polyurethane foams						
I-A-$CaCO_3$	0	0.233	Comb	1.77	0.17	0.10
	10.0	0.233	Comb	1.93	0.30	0.16
	20.1	0.233	Comb	1.89	0.37	0.20
	26.4	0.233	Comb	2.21	0.43	0.19
	38.9	0.233	Comb	3.38	0.71	0.21
	45.2	0.233	Comb	4.16	0.86	0.21
	52.4	0.233	Pyr	3.13	1.76	0.56
	52.4	0.233	Comb	4.59	1.03	0.22
	59.8	0.233	Pyr	3.43	2.79	0.81
	59.8	0.233	Comb	5.51	1.42	0.26
GM-21	45.2	0	Pyr	4.90	1.10	0.22
		0.103	Pyr	4.08	1.90	0.47
		0.139	Pyr	3.78	2.79	0.74
		0.160	Comb	4.45	0.46	0.10
		0.182	Pyr	2.34	2.78	1.19
		0.180	Comb	4.70	0.77	0.16
		0.212	Pyr	2.56	3.34	1.30
		0.213	Comb	5.51	0.83	0.15
		0.233	Pyr	2.76	2.11	0.76
		0.233	Comb	6.24	0.72	0.12
GM-23-FR	45.2	0.072	Pyr	3.55	4.38	1.23
		0.147	Pyr	2.97	3.22	1.08
		0.168	Pyr	2.36	3.62	1.53
		0.167	Comb	4.34	1.62	0.37
		0.179	Pyr	2.32	3.08	1.33
		0.177	Comb	4.01	1.43	0.36
		0.193	Pyr	2.20	3.74	1.70
		0.193	Comb	4.67	1.99	0.43
		0.207	Comb	3.43	1.51	0.44
		0.217	Pyr	2.26	2.84	1.26

TABLE 12 (Continued)

Polymer-Liquid	$\dot{q}_e''$ (kW/m²)	m_{O_2}	Mode	C_{fuel} (g/m³)	$\frac{1}{l}\ln I_o/I$ (m⁻¹)	ξ_{pyr} (m²/g)
		0.233	Comb	6.37	1.78	0.28
		0.273	Comb	6.75	1.60	0.24
GM-25	26.4	0.207	Comb	4.10	1.62	0.40
		0.214	Comb	4.29	1.75	0.41
		0.233	Comb	5.24	1.65	0.31
		0.301	Comb	8.52	1.46	0.17
	31.8	0.233	Pyr	2.12	0.93	0.44
		0.233	Comb	7.06	2.37	0.34
	38.9	0.233	Comb	7.85	2.34	0.30
	52.4	0.233	Comb	10.20	2.86	0.28
	59.8	0.233	Comb	11.19	3.34	0.30
GM-27/FR	31.8	0.233	Pyr	1.68	1.17	0.70
		0.233	Comb	6.09	2.94	0.48
	38.9	0.233	Comb	6.85	2.99	0.44
	52.4	0.233	Comb	8.56	4.13	0.48

ylenes which generate aromatic compounds are also high. This may be due not only to the differences in the physical properties of the pyrolyzate (i.e., extinction coefficient, density of the individual particles, volume-to-surface mean diameter), but also to differences in the fraction of fuel vapor converted to pyrolyzate. Some differences may also be due to temperature differences of the mixture of products and air, especially if significant amounts of high-vapor-pressure liquids are present in the pyrolyzate.

We are now extending our measurements for ξ_{pyr} where the temperature

TABLE 13
Average Values of Modified Mass Absorbency Index of Pyrolyzate Generated in the Combustion of Polymers

Polymer	ξ_{pyr} (m²/g)
Cellulose, polyoxymethylene	0.01
Polymethylmethacrylate	0.18
Polyurethane foams	0.28
Epoxy-FR-fiberglass	0.30
Polyethylene	0.35
Nylon 6/6	0.41
Polypropylene	0.48
Polyethylene foams	0.77
Aromatics (benzene, styrene, polymers, etc.)	0.98
Chlorinated polyethylenes	1.21

of the products–air mixture will be maintained constant for different polymers, i.e., increasing $\dot{V}_T$ but maintaining $\dot{m}_b$ constant.

7.4. Conclusions

For the evaluation of the fire hazard of polymers in terms of optical transmission (or visibility), ξ_{pyr} has been found to be useful. In general, ξ_{pyr} values are smaller for aliphatic-type fuels and polyurethane foams than the values for aromatic-type fuels. The chlorinated-type fuels have highest ξ_{pyr} values. Values of ξ_{pyr} are higher under the pyrolysis than under the combustion conditions. The higher the value of ξ_{pyr}, the lower is the optical transmission (or visibility) through the products.

8. Generation of Toxic Compounds

8.1. Concept

For the evaluation of the generation rates of toxic compounds, detailed qualitative and quantitative analyses of compounds in the pyrolyzate as well as non-carbon-containing compounds are needed. Our apparatus has been designed so that samples can be collected for the analyses. For qualitative and quantitative analyses techniques such as computerized gas chromatography–mass spectrometry, etc. are useful. The elemental compositions of the fuel, as well as the residue at various stages of the experiments, are also needed so that a complete balance for atoms such as C, H, O, N, Cl, etc. can be made. From the measured data for $\dot{M}_i''$, $\dot{m}_b''$, and k_i, f_i values for the compounds can be calculated (such as done for CO_2, CO, etc. in Section 5). The f_i values for the compounds can be established for a variety of combinations of $\dot{q}_e''$ and $\dot{m}_{O_2}$ values so that a generalized behavior of polymers for the generation of toxic compounds can be obtained. For the evaluation of the hazard, however, animals need to be exposed to the products at the same time as other measurements are being made in the experiments. In addition, products are required to be diluted much more than in our experiments as described in previous sections so that the temperature of the mixture of products and air is close to ambient. Note that this is also helpful for the evaluation of ξ_{pyr} values. Our apparatus has been designed so that products can be diluted to a maximum of about a hundred times their volume by fresh air; the dilution of the products can be changed as required.

In our studies we have not analyzed in detail the carbon-containing products in the pyrolyzate as well as other products and no animal experiments have yet been performed.

9. *Fire Extinction*

9.1. *Concept*

For the extinction of fire by gaseous agents, experiments can be performed with different concentrations of the agents in air entering the apparatus maintaining q''_e or m_{O_2} values constant or variable. For the application of liquid fire extinguishing agents such as water, the apparatus utilizes an arrangement where liquid droplets are applied directly to the fuel surface.[5] Fire extinction by different agents can be evaluated under a variety of experimental conditions very similar to those used for evaluating other flammability properties.

Fire extinction of fuels has not been examined in our studies so far.

Acknowledgment

The author greatly appreciates the support of Mr. R. F. Pion in performing the experiments and of Mr. J. L. Lee in data acquisition and computation. Technical discussions with Professor I. Glassman, Princeton University, Princeton, New Jersey, have proven most helpful to the study. The financial support of the Products Research Committee is appreciated.

10. *Nomenclature*

a	Sample surface area (m^2)
B_{conv}	Convective mass transfer number
B_{rad}	Radiation modified mass transfer number
C_{fuel}	Mass concentration of the fuel vapors (g/m^3)
ΔC_i	Difference between initial and final concentration of the product (volume fraction)
C_p	Specific heat (kJ/g K)
C_{pyr}	Mass concentration of the "pyrolyzate" responsible for absorption of light (g/m^3)
D	Optical density
D_{32}	Volume-to-surface mean diameter (m)
E_{eff}	"Effective" energy for ignition (kJ/m^2)
f_i	Fractional theoretical stoichiometric yield of the product or fraction of carbon in the fuel converted to the product
f_{O_2}	Fractional theoretical stoichiometric mass of oxygen consumed
H_c	Net heat of complete combustion of the fuel (kJ/g)

H_{CO}	Net heat of complete combustion of CO (kJ/g)
H_{fc}	Heat transfer to the fuel by convection from the flame per unit mass of the fuel transported from the surface to the flame (kJ/g)
h	Convective heat transfer coefficient (kW/m^2 K)
I	Optical transmission through the mixture of products and ambient air (mV)
I_0	Optical transmission through air (mV)
K	Absorption-emission coefficient (m^{-1})
$\overline{K}$	Average extinction coefficient
k	Thermal conductivity (kW/m K)
k_i	Theoretical stoichiometric mass of the product produced per unit mass of the fuel converted completely to the product
k_{O_2}	Theoretical stoichiometric mass of oxygen consumed per unit mass of the fuel for complete combustion
L_g	"Effective" heat of gasification of the fuel (kJ/g)
l	Optical path length (m)
l_m	Mean beam length (m)
$\dot{M}''_i$	Mass generation rate of the product per unit sample surface area (g/m^2 sec)
$\dot{M}''_{O_2}$	Mass depletion rate of oxygen per unit sample surface area (g/m^2 sec)
$\dot{M}''_{pyr\text{-}f}$	Mass generation rate of the pyrolyzate fraction collected on a filter paper per unit sample surface area (g/m^2 sec)
$\dot{M}''_t$	Total mass flow rate of the mixture of hot products and air through the sampling duct per unit sample surface area (g/m^2 sec)
$\dot{m}''_b$	Mass loss rate of the fuel in combustion per unit sample surface area (g/m^2 sec)
$\dot{m}_b$	Mass loss rate of the fuel in combustion (g/sec)
$\dot{m}''_c$	Critical mass loss rate of the fuel required for ignition per unit sample surface area (g/m^2 sec)
m_{O_2}	Mass fraction of oxgyen
$\dot{m}''_p$	Mass loss rate of the fuel in pyrolysis per unit sample surface area (g/m^2 sec)
p_a	Ambient pressure (atm)
Δp_d	Pressure difference between inside and outside of the sampling duct (atm)
Δp_m	Pressure difference across the measuring orifice (atm)
$\dot{Q}''_{actual}$	Actual heat release rate per unit sample surface area (kW/m^2)
$\dot{Q}''_{conv}$	Convective heat release rate per unit sample surface area (kW/m^2)
$\dot{Q}''_{rad}$	Radiative heat release rate per unit sample surface area (kW/m^2)
Q''_i	Actual, convective, and radiative heat release rates
$\dot{q}''_e$	Externally applied heat flux (kW/m^2)

$\dot{q}''_{fc}$	Flame convective heat flux to the fuel surface (kW/m^2)
$\dot{q}''_{fr}$	Flame radiative heat flux to the fuel surface (kW/m^2)
$\dot{q}''_0$	"Effective" surface reradiation loss (kW/m^2)
T_a	Ambient temperature (K)
T_d	Gas temperature in the sampling duct (K)
T_f	Flame radiation temperature (K)
T_s	Fuel surface temperature (K)
t_{ig}	Ignition time (sec)
Δt	Time during which pyrolyzate fraction is collected (sec)
$\dot{V}_T$	Total volumetric flow rate of the mixture of products and air (m^3/sec)
$\dot{v}$	Volumetric flow rate of the mixture of products and air through the filter paper (m^3/sec)
Δw	Mass of the pyrolyzate fraction collected on the filter paper (g)
α	Surface absorptance
δ_i	Ratio of the density of the product to the density of air
ρ	Density (g/m^3)
ϵ	Surface emissivity
σ	Stefan–Boltzmann constant (kW/m^2 K^4)
ϕ	H_{fc}/H_c
ξ_{pyr}	$(1/C_{fuel}\, l) \ln(I_0/I)$
χ_{actual}	Combustion efficiency of the fuel vapor
χ_{conv}	Convective fraction of the theoretical stoichiometric heat release rate for complete combustion of fuel vapor
χ_{rad}	Radiative fraction of the theoretical stoichiometric heat release rate for complete combustion of fuel vapor

11. List of Components for Experimental Apparatus

1. Calibrated flow meters—Matheson Gas Products, Lyndhurst, New Jersey
2. Steel gas mixer, 0.076-m ID, 0.24 m in length
3. Glass beads (1.5 mm in diameter)
4. Flow meter—Model 2012, Thermo System Inc., St. Paul, Minnesota
5. Continuous oxygen paramagnetic analyzers: Models F-3 and 755, Beckman Instruments, Inc., Fullerton, California
6. Pumps
7. Water-cooled load cell, Model No. UC3, Statham Instruments, Inc., Oxnard, California
8. Uni-strut frame
9. Quartz tube, 0.61 m in length and 0.17 m in diameter (internal)

10. Aluminum sample holder, ~0.10 m in diameter (internal) and ~0.02–0.05 m deep
11. Sample, maximum diameter 0.10 m, maximum thickness 0.02–0.05 m
12. Pilot flame
13. High-density radiant heaters (four, only two are shown) Model 5208, Research Inc., Minneapolis, Minnesota
14. 0.0016-m stainless steel funnel, internal diameter of stem hole, 0.0310 m
15. 0.00018-m stainless steel duct, 0.10-m ID, 0.61 m long
16. Fiberglass insulation, 0.1 m thick, 0.61 m long
17. Thermocouples—type TAS, exposed junction Chromel-Alumel, United Sensor and Control Corporation, Watertown, Massachusetts, and Chromel-Alumel Ribbon Thermocouples, 36FF, ISA type K, Nanmac Corporation, Framingham, Massachusetts
18. 0.001-m ID stainless steel particulate probe
19. Aerosol filter holders, Model XX50-04700, containing two filters—Mitex (Teflon), type LS, mean pore size 5.0 μm and MR-Millipore (mixed ester of cellulose), type DA, mean pore size 0.65 μm. Millipore Corporation, Bedford, Massachusetts
20. Limiting orifice, Model XX50-000-00, Millipore Corporation, Bedford, Massachusetts
21. Probe attached to a cell (not shown) to measure the major toxic compounds generated in the fire
22. 0.001-m ID stainless steel gas sampling probe
23. 0.05077-m orifice plate, 0.0032-m-thick stainless steel
24. Inclined-vertical manometer, Model 424-5, Dwyer Instruments Inc., Michigan City, Indiana
25. Traps (two, only one is shown) for condensible compounds immersed in a cooling bath
26. Drier (anhydrous $CaSO_4$), W. A. Hammond Drierite Company, Xenia, Ohio
27. Continuous CO IR analyzer, Model 315A, Beckman Instruments, Inc., Fullerton, California
28. Continuous CO_2 IR analyzer, Model 315A, Beckman Instruments, Inc., Fullerton, California
29. Continuous hydrocarbon flame-ionization analyzer, Model 400, Beckman Instruments, Inc., Fullerton, California
30. Light source
31. Photocell
32. 0.025-m-thick steel box, 0.61 m long, 0.30 m wide, and 0.30 m high
33. Blower, Model HP 33, Clements National Company, Chicago, Illinois

12. *References*

1. H. S. Carslaw and J. C. Jaeger, *Conduction of Heat in Solids*, 2nd ed., Chap. II, p. 75, Oxford University Press, London (1959).
2. D. J. Rasbash, Relevance of fire point theory to the assessment of fire behavior of combustible materials, in *International Symposium on Fire Safety of Combustible Materials*, p. 169, University of Edinburgh, Center for Industrial Consultancy and Liaison, Edinburgh (1975).
3. D. J. Rasbash, Theory in the evaluation of fire properties of combustible materials, in *5th International Fire Protection Seminar*, p. 113, Karlsruhe, West Germany (1976).
4. A. Tewarson, paper submitted to *Combustion and Flame.*
5. A. Tewarson and R. F. Pion, *Combust. Flame*, **26**, 85 (1976).
6. S. L. Madorsky, *Thermal Degradation of Organic Polymers*, Wiley–Interscience, New York (1964).
7. D. B. Spalding, The Combustion of Liquid Fuels, in *Fourth Symposium (International) on Combustion*, p. 847, The Combustion Institute, Pittsburgh (1953).
8. I. Glassman, The Mass Burning Rate of Polymeric Materials, The Center for Environmental Studies, Princeton University, Princeton, New Jersey, Report No. 23, (June 1975).
9. A. Tewarson, J. L. Lee, and R. F. Pion, The Influence of Oxygen Concentration on Fuel Parameters for Fire Modeling, 18th Symposium (International) on Combustion, The Combustion Institute, Pittsburgh, p. 563 (1981).
10. H. C. Kung, private communication (November 1978).
11. J. de Ris, Fire Radiation—A Review, *17th Symposium (Internal) on Combustion*, The Combustion Institute, Pittsburgh, p. 1003, (1979).
12. R. L. Alpert, J. de Ris, A. T. Modak, M. K. Mathews, G. H. Markstein, L. Orloff, R. I. Land, E. Farren, J. C. Polo, D. S. Mann, F. B. Kiley, and L. Crudup, Influence of Enclosures on Fire Growth—Vol. I Test Data, Factory Mutual Research Corporation, Norwood, Massachusetts, Technical Report No. 0A0R2.BU-7, (August 1977).
13. W. P. Chien and J. D. Seader, Smoke Measurement in a Modified NBS Smoke-Density Chamber, Flammability Research Center/Department of Chemical Engineering, University of Utah, Salt Lake City, Utah, Report No. FRC/UU-44, UTEC 75-055, (April 1975).

4

Flammability Evaluation Methods for Textiles

John F. Krasny

1. *Introduction*

This chapter will discuss flammability evaluation of fabrics as well as of end-use items containing fabrics, such as apparel, curtains and draperies, tents, mattresses, upholstered furniture, etc.

Flammability evaluation methods can be classified in various ways. Examples of possible classifications follow.

Tests, Standards, and Simulation or Modeling Experiments. Test methods usually describe test equipment and procedures, and generally produce numerical results, e.g., ignition time or char length. They are useful for comparing various textile items and flame-retardant (FR) treatments. In addition to test methods, standards usually also contain recommendations for sampling, record keeping, laundering, drycleaning, weathering, and aging before testing, as well as pass–fail criteria. Modeling experiments consist of burning of curtains, chairs, or bedding in rooms, carpets in rooms and corridors, garments on mannequins, etc. They provide valuable information on the interaction of flammable items with their surroundings, and on flame spread and heat, smoke, and toxic pyrolysis product evolution. Such experiments, while generally costly, are often indispensible for the development of meaningful test methods. The best test methods are those which

John F. Krasny • Center for Fire Research, National Bureau of Standards, Washington, D.C. 20234.

simulate the occurrences during such full-scale experiments or real-life fire accidents as closely as possible, but are still not too complex for routine use.

Issuing Organization. In many countries, tests and standards are developed by agencies which represent both industry and government, e.g., the German DIN and British Standards Institution. Other tests and standards are developed by technical organizations such as the American Society for Testing Materials (ASTM), the National Fire Protection Association (NFPA), and others. Standards in the U.S. are also issued by many federal, state, and local agencies. Among the federal agencies are the following: the Consumer Products Safety Commission (CPSC), which is responsible for consumer goods; the Department of Transportation (aircraft, buses, trains, cars); the Department of Health, Education, and Welfare, the Veterans Administration, and the Social Security Administration (health care facilities under their respective jurisdictions); the General Services Administration (federal buildings), and the Department of Defense. State standards exist in California for mattresses and upholstered furniture, in Minnesota for tents; and in most states or cities for curtains in public buildings. A useful compilation of federal standards covering textile items has been published by the American Association of Textile Chemists and Colorists in 1975.[1]

Type of Protection. Some evaluation methods are designed to characterize the behavior of textiles when exposed to a small ignition source, others to determine the contribution to the heat evolution and progress of the fire in a conflagration. Still other evaluation methods are primarily used for research, e.g., OI, TGA, DTA, etc. In the last few years, it has been learned that smoldering ignition, such as from a cigarette, can cause very different phenomena in textiles than flame ignition, and tests for resistance to this type of ignition have been developed. Finally, there are evaluation methods for the insulative or protective values of textiles in fire situations, for fabrics used in, e.g., firefighter's or steel worker's garments.

Component or assembly testing: in many cases, only the fabric is tested, and assumed to represent the end-use item, such as curtains or garments. However, the Children's Sleepwear standards[2] require that besides the fabric, parts of the garments containing seams and trim be tested. Similarly, the U.S.[3] and Canadian[4] mattress tests require testing of the cover fabrics and filling of the mattresses assembled in exactly the same manner as in a mattress product line, because it has been shown that separate testing of component materials is not a reliable guide to the flammability of the assemblies.

Criteria. Criteria used in textile flammability evaluations of items which burn are ease of ignition, flame spread rate, evolution of heat, smoke, and toxic materials, and ease of extinction. For test methods designed to determine self-extinguishment of textiles after the ignition source is removed, such criteria as char length, afterflame and afterglow time, and presence of burning or nonburning melt drip are recorded.

2. *Evaluation Methods for Fabrics Which Are Expected to Self-Extinguish*

Self-extinguishing fabrics are those which stop burning after removal of an ignition source. Such fabrics are often required clothing for persons which cannot be expected to respond effectively when they find their clothing on fire, such as children and hospital patients, and for clothing of workers routinely exposed to heat and fire. They are also used for curtains and other decorative materials in public buildings, and for tents.

The usual criteria for self-extinguishment are char length, afterflame and afterglow time, and presence of melt drip.

Char length is the length of the fabric area that was destroyed by the flames, measured from the point of flame application. The destruction can be due to charring or, in thermoplastic fabrics, to melting. The length is determined by hooking a weight to one side of the undamaged specimen area (where it was held in a frame), and lifting the other side. The length of the torn or melted area is the char length. The weight increases with fabric weight; e.g., for the children's sleepwear tests,(2) see Table 1. Other test methods use similar relationships.

Afterflame time is the time the flame continues to burn after the ignition flame is removed, and is determined visually. Afterglow time is the time glow persists on the specimen after it has ceased to flame.

Residual flame time is the time burning fragments falling from the fabric burn on the bottom of the cabinet. This time is often measured in various test methods, but is not always included in the pass-fail criteria. In apparel, such fragments—often called melt drip—have a limited injury potential, but they can be the source of ignition for other garments, or in the case of curtains, furnishings.

TABLE 1

Fabric weight (g/m^2)	Weight (g)
Less than 101	54.4
101–207	113.4
207–338	226.8
More than 338	340.2

There are numerous tests for self-extinguishing fabrics. Most are quite similar, and specify vertical specimens ignited at the bottom edge. This method works well for determining the efficacy of FR treatments on cellulosic or wool fabrics, because they burn upwards. Thermoplastic fabrics, however, tend to shrink and melt ahead of the flame, and to ablate as melt drip or in the form of fabric pieces. They often do not burn when ignited at the bottom edge though they may burn when ignited on their surface. Such burning is often sidewards and downwards as well as upwards.

Ignition is generally by methane, propane, butane, or Mathieson B gas (55% hydrogen, 24% methane, 3% ethane, and 18% carbon monoxide) burners, though candles and matches are also used. The burners range from simple tubes through Bunsen burners to more sophisticated, precision-made, premixed flame burners.

In most cases, the specimens are held by U-shaped frames; for some fabrics, suspension in frames affects the results, e.g., woven aramid (Nomex), modacrylic, and thermoplastic fabrics. Such fabrics often show an occasional long char length which does not occur when the fabrics are suspended freely, as in the form of a dress on a mannequin.[5,6] Other fabrics show longer char length if they are not held tightly in the frames. Pin frames combined with wider specimens than used in most U-shaped metal frames are specified in many tests, to overcome the frame effects.

Most tests are conducted in a cabinet, with vent openings at the top and bottom.

Preconditioning of samples is required in most test methods. Conditions vary from the standard textile conditions, 21°C and 65% relative humidity, to oven drying at 105°C followed by cooling in a desiccator for various periods of time. A maximum time from removal from the desiccator to ignition of the specimens is usually prescribed. Only one country, Australia, has recognized the effect of altitude on flammability test results[7] and specified a minimum air pressure during testing.

Other preparations of specimens include brushing of raised or pile fabrics, so the nap or pile will be erect, and laundering, drycleaning, aging, leaching, or weathering of the specimens before testing.

Most tests undergo an interlaboratory evaluation before they are used in regulatory or voluntary standards, to assure that they are reproducible and repeatable. Frequently, such interlaboratory evaluations lead to changes or clarifications of the test methods. Completion of at least one interlaboratory evaluation (round-robin) is generally required before adoption of a standard by the International Standards Organization (ISO), the CPSC, ASTM, and others.

In the following, test parameters will be given for many tests. In general, specimen size is given because of the above-mentioned effect of frames which tends to diminish with increasing specimen size, and because in fabrics which burn, flames tend to accelerate and small specimens cannot be considered as good predictors of the behavior of garments or tents as large specimens. Other items which affect test results, such as preconditioning of the specimens, ignition sources, and flammability criteria are also listed.

2.1. Vertical Tests

The U.S. Children's Sleepwear Tests[2] apply to fabrics and garments, including areas which contain seams and trim. Specimen size is 254 × 89 mm. The specimens are suspended in frames, oven dried at 105°C, and cooled. Ignition is at the bottom edge with a 38-mm methane flame for 3 sec. To pass, the average char length must not exceed 178 mm, and no more than one specimen can have a char length of 254 mm (with one 254-mm char length, retesting is permitted).

Children's sleepwear was singled out for stringent flammability testing not only in the U.S. but also in Australia and Great Britain because statistics indicated that children were involved in a larger number of fire accidents than would be expected from their percentage of the population.[8] A substantial part of these accidents occur while children wear nightclothes.

The children's sleepwear test for younger children, Part 1615 (FF 3-71), originally contained a requirement that melt drip must not burn more than 10 sec on the floor of the cabinet, while the standard for the larger sizes did not contain this requirement. Many polyester and nylon fabrics, particularly in seamed areas, could not consistently pass this requirement. A flame-retardant, tris(2,3-dibromopropyl)phosphate, was widely used on polyester fabrics and made it easy to pass the test. It was later suspected to be carcinogenic, and the CPSC removed the residual flame time requirement from FF 3-71.[9] This ruling made it possible to use untreated polyester and nylon fabrics, primarily knits, for children's sleepwear.

A frequent objection to this and other tests has been the oven drying requirement. This requirement necessitates higher concentrations of FR finishes on cellulosic and wool fabrics than when the specimens are conditioned at higher relative humidities. However, oven drying was justified by a study which showed that garments which are approximately 30 mm from a child's body are at equilibrium with the ambient conditions in the room, and are not affected by body moisture.[10] Ambient conditions in houses can, of course, be very dry in many regions in the U.S. In addition, it was shown that fabric samples held in front of space heaters for 5–10 min reach oven dry conditions.

Since seam testing is required by the children's sleepwear standards, a study of the effect of seam parameters on char length was conducted.[11] The most important factor in reducing char length of seams in thermoplastic fabrics is the use of a sewing thread with similar melting characteristics as the base fabric.

The children's sleepwear standards had considerable impact on the fiber distribution in this market. Cotton, which prevailed in this market before the standards, has only a small share now, and polyester, nylon, modacrylics, and matrix fibers and their blends predominate. The standard has been found to be cost effective[12] even though the cost of the garments in this study was based on the price of FR cotton, which is higher than that of polyester and nylon garments, and the cost of permanent disfigurement and psychological damage to burned children was not considered. The two standards have apparently greatly reduced the number of children in burn wards.[13,14]

The "Textile Test Methods" of the "Federal Test Method Standard" contain several flammability tests, starting with No. 5900. Method 5903[15] is similar to the children's sleepwear tests; it covers self-extinguishing fabrics except pile fabrics. Specimen size is 305 × 70 mm, conditioning at 21°C and 65% relative humidity. Specimens are suspended in a U-frame, and ignited at the bottom edge with a 38-mm Matheson gas B flame. Criteria are char length, and afterflame and afterglow time. This test is used in the following standards: CPAI 84[16] for tents and FAA regulation Part 25.853[17] for floor coverings, draperies, upholstery, and other decorative fabrics in aircraft.

Method 5904[18] specifies a specimen size approximately 127 × 51 mm, a vertical frame but, unlike the previously described tests, no cabinet (it is a field test). Conditions are ambient, ignition is by candle for 12 sec at the bottom edge. The same criteria as in method 5903 are used.

Method 5905[19] was developed primarily for high-temperature resistant fabrics to supplement method 5903. Conditioning is at 21°C and 65% relative humidity, and the specimen size is again 305 × 70 mm. The specimen is held with tongs in a 76-mm butane flame so that its lower edge is 38 mm above the burner edge for 12 sec. If the specimen shrinks, the specified distance from the burner edge is maintained during this time. Afterflame is timed. When afterflaming ceases, the specimen is again held in the flame for 12 sec. The term *ignition, self-extinguishing* is used for specimens which extinguish before the entire length of the specimen is consumed. *Ignition-resistant* specimens do not flame during exposure or after withdrawal from the flame. Heat shrinkage and melt drip are also noted.

NFPA 701-76[20] is intended for curtains, furnishings, and scenery in places of public assembly, hotels, hospitals, for tents, awnings, tarpaulins, and certain work clothing. This test is used widely by local code authorities. It consists of a small- and a large-scale test. The small-scale test uses a 254 × 70 mm specimen, in a vertical frame. Conditioning is in an oven at 60°C. Ignition is by a 38-mm natural gas flame at the bottom edge for 12 sec. Pass–fail criteria are a maximum 2-sec afterflame time and char lengths which vary with fabric weight from 165 mm for fabrics 195 g/m^2 or less to 115 mm for fabrics exceeding 330 g/m^2 (this is based on the finding that char length of properly treated FR fabrics decreases with weight for any flame exposure). NFPA

701 also describes a small-scale field test using a freely suspended specimen and a match.

The NFPA 701 large-scale test specifies a specimen size of 2970 × 127 mm for flat specimens, 2970 × 635 mm for specimens which are arranged in four folds simulating curtains. There is no specimen frame. Specimens are conditioned in an oven at 60° C. Ignition is by means of a 280-mm natural gas flame for 12 sec at the bottom edge. Pass–fail criteria are a maximum 254-mm char length above the flame tip for flat, 890 mm for folded specimens; no flaming melt drip is permitted.

Many local jurisdictions use NFPA 701 for public buildings, others have developed their own methods. Most of these use gas flames or matches, framed or free-hanging specimens, and char length, afteflame and afterglow time criteria. The results should not differ greatly from those obtained by method 5903 or the small-scale NFPA tests.

The "Semirestraint" test, ASTM D 3659[21] was developed to overcome the problems associated with specimen frames in the children's sleepwear and similar tests, especially the unpredictable, occasional excessive char length results. It specifies the children's sleepwear test cabinet, oven drying at 105° C, and desiccator cooling. However, the 381 × 152-mm specimen is suspended from a horizontal rod. Its bottom edges are connected to two restraining chains which allow some movement and shrinkage of the specimens. Ignition is by means of a fan-shaped, 38 mm, methane flame, at the bottom edge for 3 sec. Criteria are burn time, weight loss, and destroyed area. In another version, the specimens are ignited by flames from two burners located on either side of the center of the specimen. A number of papers have been written about this test[5,6] and the correlation of the results with burns of garments on mannequins. An interlaboratory study of the test was conducted, and repeatability and reproducibility were reported to be satisfactory.

The vertical self-extinguishment test[22] used in West Germany is typical of European practices. Specimens are 340 × 75 mm, held in frames, conditioned at 21°C and 65% relative humidity. Ignition is by means of a rather complicated propane burner with a 40-mm premixed flame, at the bottom edge, for 3 and 15 sec. Criteria are char length, afterflame and afterglow time, and existence of melt drip. A separate standard[23] covers pass–fail criteria for curtains.

Canada[24] specifies 315 × 50 mm specimens, held on top only, and ignition of the bottom of the specimen by a 40-mm Bunsen burner flame for 12 sec. No preconditioning of the samples is mentioned. Test criteria are occurrence of flashing, afterflame and afterglow time, char length, and occurrence of melting, shrinking, or dripping. Fabrics which do not pass this test are evaluated by means of another vertical test.[25] The specimens, 760 × 50 mm, are oven dried, held by a clamp at the top only, and a 15-mm microburner flame is applied to their bottom edge until ignited. Trip threads are positioned at 25 and 675 mm above the bottom edge of the specimen, and the time between the burning of the lower and upper thread is reported.

2.2. Horizontal Tests

Most horizontal methods for self-extinguishment were designed for such items as mattresses and carpets, and will be discussed later. Method 5600,[26] however, is designed for FR fabrics and open-weave, untreated fabrics such as netting. Specimen size is 178 × 254 mm, conditioning in an oven at 60°C; 0.3 ml of ethyl alcohol is ignited under the specimen, which is suspended in a frame. Criteria are the char length (the greatest distance across the hole) and afterflame time. Horizontal specimens are also used in ignition time tests for fabrics which burn, as discussed in Section 5.

3. Standards for Both Self-Extinguishing and Flammable Fabrics

The Australian and British standards include tests for both self-extinguishing and flammable fabrics. The Australian Standard AS 1176[27] contains three test methods: Part 1, ease of ignition; Part 2, burning time and heat output; and Part 3, surface burning times for pile fabrics. The results of these tests are used in conjunction with standards for fit for children's nightwear and for labeling. Conditioning for Parts 1 and 2 is at 20°C and 65% relative humidity, Part 3 requires oven drying. The tests must be conducted at a barometric pressure of not less than 1000 mbar.

In Part 1, the specimen holder consists of two metal plates with seven holes, 40 mm in diameter and 50 mm apart. The holder is horizontal, and is placed, with one hole at a time, above the ignition source, a 25-mm hydrogen flame. Specimens are taken from fabrics as well as garments. Ignition time is determined by preliminary testing, and then at least 20 specimens are exposed so that ten ignite and ten do not ignite. Ignition is defined as appearance of a scorch mark with a dimension of at least 20 mm in any direction on the upper side of the specimen.

Part 2 measures burning time and heat output. Specimens are cut from fabrics and garments, and are 700 × 40 mm. They are stitched with two rows of glass thread, to avoid ablation of thermoplastic fabrics. They are then mounted on vertical pin frames. Ignition is at the bottom edge with a seven-jet hydrogen fan burner using a 40-mm flame for 5 sec. The time between the burning through of two No. 50 cotton sewing threads held near the fabric surface (trip threads), located 100 and 600 mm from the bottom end of the specimen, is measured. Heat output is determined by measuring the temperature rise of a 150-g, 160-mm-long, 12-mm-diameter copper rod mounted near the top of the cabinet.

Part 3 measures the flame spread rate of napped or pile fabrics. Specimens are 150 × 75 mm, and are conditioned in an oven at 105°C. The specimens are attached to a vertical steel plate, and ignited by a hypodermic needle burner using hydrogen, with a 50-mm flame, for 0.5 sec. The point of ignition is high on the specimen, and the fabric "shall be reported as not having surface burning properties" if the flame does not reach a reference mark 75 mm below the point of ignition.

AS 1176 is supplemented by a rather complex system of standards covering labels, sizing, and placing garments into various categories of hazard.[28-31] AS 1248[28] defines "fabrics for domestic apparel of the low fire hazard types" as follows: ignition time, no less than 6 sec; burning time, not less than 18 sec average, with no specimen less than 16 sec (unless the specimens self-extinguish within 500 mm from the point of ignition in any direction); and the surface burning time on both surfaces shall be not less than 10 sec. Fabrics conforming to these specifications can be used in "children's night clothing having low fire hazard," covered by Part 1 of AS 1249,[29] which also specifies that the garments must be fairly tightly fitting. The definition of the required fit is based on garment measurements shown in yet another standard, AS 1182.[30] Part 2 of AS 1249, "children's night clothes designed to reduce fire hazard," permits garments from any fabric but they must conform to the above sizing specifications. In addition, fabrics which melt and drip must not be mixed with fabrics which do not in any one garment. Part 3 of AS 1249 refers to garments which do not conform to the requirements of Parts 1 and 2, and which are limited to relatively short garments. If they are made from pile fabrics, the burning time must exceed 10 sec. AS 1989[31] reviews the above classifications and specifies labels.

Many of the concepts used in Australian standards are unique. They are explained in the backup reports for the standards.[7,32] The hydrogen burner was chosen to obtain a stable, less severe flame than the usual gas flames. The horizontal position for the ignition test is admittedly arbitrary but has been used for the same purpose by many other test methods. Tests were conducted on the same fabrics at various altitudes and showed different results; this led to the barometric pressure specification. Heat output was shown to show similar patterns as burning times, but was less statistically reliable. Like the U.S.,[10] Australia has found that loose garments are more likely to ignite and to inflict severe injuries than tightly fitting garments, and garment fit was considered in the standards system.

The British Standard, BS 5438[33] also has several parts. In all cases, the specimens are vertical, suspended on pin frames, conditioned at 15–25°C and 55%–65% relative humidity. A premixed, horizontal, 45-mm butane flame is the ignition source. Unlike most other tests, no cabinet is specified.

The ease-of-ignition test specimen is 220 × 170 mm. The point of ignition is about 45 mm above the lower edge. Ignition is assumed to occur if the fabric burns for 1 sec after the ignition flame is stopped.

In test 2, the flame is applied to the same size specimen for 10 sec. The following are observed: afterflame and afterglow time, flaming debris, the maximum dimensions across holes in the fabric, and whether flame or glow reaches the upper and side parts of the frame. If the fabric self-extinguishes but forms a hole, additional specimens are tested with 2-sec ignition times.

Part 3 measures the vertical and horizontal flame spread rate by means of horizontal and vertical trip threads, located 75 mm on either side of the ignition point and 300 and 600 mm above it. Specimen size is 670 × 170 mm. In addition to the time to burn through all of the four trip threads, afterglow and afterflame times and flaming debris are noted.

The standard has several unique features, such as the already mentioned lack of a cabinet and the vertical ignition test. There is also a warning paragraph: "The results may not apply to situations where there is restricted air supply or prolonged exposure to large sources of intense heat as in conflagration." In addition, there is information on handling of "outlying" values for each test, based perhaps more on experience than statistics. Finally, no pass–fail criteria are given. BA 5438 replaces two earlier standards, which formed the bases for two performance standards which also have been declared obsolete. Performance standards based on BS 5438 are under development.

4. *Test Methods for Flammable Fabrics*

4.1. *Flame Spread Rate Tests*

Most tests designed for fabrics which are expected to burn use a flame spread rate (burn time) criterion. The flame spread rate is generally determined from the time after ignition until a cotton trip thread is burned through. Specimen sizes in these tests vary widely. This can affect the results since flame spread rate often increases with increasing distance from the point of ignition.[34,35] Specimen orientation is mostly 45° or vertical though other angles have been used. For many fabrics, the ranking is similar regardless of orientation, but others, mostly thermoplastic fabrics, often burn more readily at angles other than vertical. Some of the methods were designed to evaluate fabrics which burn as well as self-extinguishing fabrics, and the criteria are then flame spread rate or char length and afterflame and afterglow time.

The measurement of flame spread rate is based on the assumption that high flame spread rates relate to high hazard. A British study[36] showed already in 1958 that this assumption is questionable. A more extensive U.S. investigation confirmed this finding. The flame spread rate of remainders of about 700 garments salvaged from apparel burn accidents showed no relationship to the size of the burn injuries of the victims of these accidents.[37] Similarly, full-scale fires of curtains showed considerably more involvement of wall and ceiling materials when heavy curtains burned slowly, with considerable heat evolution, than when lightweight curtains burned rapidly.[38] Nevertheless, the majority of efforts to rank the relative hazard from burning fabrics is still based on flame spread rate tests, though other tests for this purpose have been developed.

1. 45° Tests. The U.S. standard for the flammability of clothing textiles, CS 191-53,[39] was specifically designed to stop frequent accidents involving fuzzy rayon sweaters (torch sweaters) and children's cowboy chaps which occurred in the late 1940s. It served this purpose well; it also sometimes fails cotton flannels, stiffened nylon netting, and light, open, cellulosic fabrics.

Specimens measuring 152 × 51 mm are placed in frames, conditioned in an oven at 105°C, and cooled. Napped specimens are brushed so the pile is erect. The specimen is mounted in a cabinet at 45° to the vertical. A 16-mm butane flame impinges on the upper surface of the specimen 25 mm from its lower end, for 1 sec. The time until a cotton trip cord 127 mm from the point of ignition is burned through is measured.

Fabrics with such burn times of less than 4 sec cannot be sold in the United States. Furthermore, fabric classes were established. For Class 1, normal flammability, the burn time must be 4 sec or more for fabrics without nap or pile, and 7 sec for those with nap, unless the flame is confined to the nap or pile and does not burn the substrate fabric. Intermediate flammability, Class 2 fabrics, are napped or pile fabrics with burn times of 4–7 sec, and involvement of the base fabrics. Finally, Class 3, rapid and intense burning, are those with burn times of less than 4 sec, with burning of the substrate in the case of napped and pile fabrics.

The provisions of NFPA 702-1975[40] are similar to those of CS 191-53 except that ignition is at the bottom edge of plain fabrics, and is continued until the specimens start to burn (forced ignition) rather than for 1 sec. For pile fabrics, the CS 191-53 ignition procedure is used. Federal Test Method 5908[41] also follows this procedure. Similarly worded ASTM and AATCC methods have been discontinued.

Canada has two 45° tests, one based on CS 191-53[42] but not using the classification system, the other quite different, and establishing a classification system.[43] The latter method specifies 180 × 150-mm specimens oven dried at 105°C. A cup with 0.3 ml of ethanol (which burns approximately 50 sec) is placed below the specimen. Fabrics are in the "flame-resistant" class if the char or scorch does not reach the top edge of the specimen, and if there is no afterflame or afterglow after cessation of the ethanol flame. They are "moderately flame resistant" if the char or scorch does not extend more than 50 mm across the top edge, and if the afterflame and afterglow stop within 15 sec after cessation of the flame. Other fabrics are in the "not-flame-resistant" class.

West Germany's DIN 54335[44] requires a 600 × 100-mm specimen, conditioned at 21°C and 65% relative humidity. The 40-mm premixed butane flame is applied to the bottom edge of the specimen for 15 sec. Flame spread rate, afterglow and afterflame time, and melting are observed.

Japanese testing procedure includes 45° testing of thermoplastic fibers.[45] The fibers are inserted into a cylindrical steel coil, and ignited near the bottom by a microburner. If they extinguish, they are again ignited at a higher point, until they have burned the entire 90-mm specimen length. Fibers which have to be ignited at least three times are considered flame retardant.

A tentative ASTM method, ASTM D 3411-75T,[46] combines a test for ease of ignition with a 45° flame spread rate test. For ignition testing, 114 × 89-mm specimens are conditioned at 21°C and 65% relative humidity, and exposed in a frame inclined at 62½° to the horizontal. A 15-mm microburner, methane flame is impinged on the specimens, and the time to ignition determined by repeated trials. The flame spread rate test is conducted on 279 × 64-mm specimens dried at 105°C. The same flame is applied to the bottom edge of the specimens until the test is completed or for 15 sec. The time of flame travel between 111 and 263 mm from the lower specimen edge is determined.

2. *Vertical Tests*. The Australian[27] and British[33] vertical flame spread rate test methods, which are part of a standard combining other tests, have been discussed earlier. The U.S. children's sleepwear test equipment can be used for flame spread rate measurements by adding a trip thread at the top of the frame.[47] This method was developed in the laboratory of an insurance company to provide its textile industry clients with a means to identify fabrics which would be most likely to cause burn accidents.

The two Canadian vertical flammability test methods[24,25] have already been described.

3. *Other Specimen Orientations*. Method 5910[48] specifies a 30° angle to the horizontal, a wire grill frame, conditioning in an oven at 60°C, and ignition by book match for 5 sec on an end protruding 3.2 mm from the frame and pointing downward. The time to burn the 153-mm framed part of the specimen, or the char length and afterflame time for fabrics which extinguish, is noted. The method is recommended for thermoplastic materials.

Method 5906[49] measures horizontal burning rate. Specimen size is 318 × 115 mm. The specimen is conditioned in an oven at 60°C, and the same temperature is maintained in the cabinet by strip heaters. Ignition is with a 38-mm Bunsen burner flame, and the time of 254-mm flame travel is reported.

Motor Vehicle Standard No. 302[50] covers the flammability of interior materials of passenger cars, trucks, and buses. Specimen size is 356 × 102 mm; conditioning at 21°C and 65% relative humidity. A 38-mm natural gas,

Bunsen burner flame is applied to the edge of the horizontal specimen for 15 sec. The time it takes for the flame to travel from a point 38 mm from the ignited edge to a point 292 mm from this edge is measured, or the time until extinguishment if the flame does not travel this distance. The flame spread rate measured in this manner must not exceed 102 mm/min, or the material must self-extinguish within 60 sec and within 51 mm from the point of ignition.

The West German DIN 54333, horizontal flame spread test[51] specifies a 350 × 100-mm specimen, conditioning at 20°C and 65% relative humidity, and the 40-mm premixed propane flame used in all West German standards, applied to the specimen edge for 15 sec. Char length or flame spread rate are reported.

Some years ago a semicircular flammability tester became popular, and one version became a West German standard.[52,53] The specimens were clamped to a 38- or 50-mm-wide semicircular frame with a 356-mm diameter. They were ignited by alcohol or gas flames. The time to burn around the track, or the angle to which self-extinguishing specimens burned, were noted. For a number of cellulosic materials, the vertical flame spread rate was found to be proportional to the square root of the distance burned on the semicircular tester. The advantage of this method was that both self-extinguishing and flammable materials could be evaluated, and that less ablation of thermoplastic materials was encountered than in vertical test methods.

4. Discussion of Flame Spread Rate Test Results. Below is a summary of some of the numerous papers dealing with the development of flame spread tests and their correlation with each other and other tests. Emphasis is on recent papers, which provide references to earlier work.

A mathematical analysis of the speed of propagation of flame on fabrics explains its dependence on fabric weight, but also casts doubt on the relevance of flame spread measurements with respect to burn hazard.[36] On the other hand, a more recent analysis came to the conclusion that not the average flame spread rate but the upward acceleration of the flame front is important.[54] As the average flame spread rate, this acceleration decreases with increasing fabric weight for most cellulose-containing fabrics, and it becomes negligible for fabrics weighing more than 233 g/m^2. (As will be discussed in Section 7, heavy fabrics are often hard to extinguish, and they produce more heat per unit area: these two factors can undoubtedly affect the outcome of a clothing fire.)

The major deficiencies of the CS 191-53 standard are its unrealistic angle, short ignition time, and low pass–fail requirements. It is compared with the NFPA 702 method, which is almost identical but for forced ignition at the fabric edge.[55] Edge ignition resulted in shorter ignition times, and many fabrics which did not ignite with surface ignition burned readily with edge ignition. If the flame is applied to the upper surface of cellulosic fabrics, it is not likely to break through the lower surface with fabrics weighing more than 180 g/m^2. Edge ignition causes burning on both sides from the very start, and consequently higher flame spread rates. Breakthrough of the flame to the bottom side occurs earlier in cellulosic–thermoplastic blend fabrics than in all-cotton fabrics. The data produced by the forced, edge ignition NFPA 702 method have a linear relationship with fabric weight, while the relationship between fabric weight and burn time with forced surface ignition is parabolic. Flannels, corduroys, and velveteen burned similarly to flat fabrics in these tests.

Comparison of the flame spread rates of horizontal, 45°, and vertical specimens ranked fabrics generally in the same manner, except for acrylic and acetate fabrics which burned quite slowly in the horizontal position.[34,56] The flame spread rates for thermoplastic fabrics tend to be low because of melting ahead of the flame, but even small percentages of finish can prevent the melting material from moving downward. Glass fiber stitching along the longer specimen dimension obviates this ablation; when it is used, the inverse relationship between fabric weight and flame spread rate is observed for thermoplastics as well as for cellulosics.[57] However, when tested in garments on mannequins, or as curtains, thermoplastics often burn slowly, with little heat evolution, or self-extinguish. Only when the thermoplastic is in contact with another fabric, e.g., a cellulosic underwear item, does it burn like the glass thread stitched specimens.

A summary of the work which led to the development of the British standard[58] indicates that a small gas flame is more likely to ignite thermoplastic and some FR-treated cotton fabrics than a large flame. For most other materials, the ignition source was unimportant. Vertical flame spread was faster on wide than on narrow specimens. Some fabrics, such as acrylics and acetate, have considerable lateral flame spread.[59] Double layers of fabrics burn at about the same rate as the faster single layer. Experiments with cylindrical specimens gave similar results as relatively wide, flat specimens on pin frames.

The Textile Research Institute conducted two types of flame spread experiments as part of its textile flammability program. The TRI Fabric Flammability Analyzer determines a steady state rate of flame propagation.[60] A specimen mounted on a circular frame is moved so that the flame always remains at the same position in space, e.g., 3, 6, 9, or 12 o'clock. The oxygen/nitrogen ratio in the cabinet can be regulated. The results have been interpreted mostly in terms of the relationship between oxygen concentration and steady flame spread rate, and will be discussed in the chapter on extinguishability.

The other TRI work consisted of burning unrestrained fabric strips.[61] In this arrangement, the mass burning rates per unit area density of the fabrics were low for nylon and polyester, intermediate for acetate, triacetate, wool and acrylic, and high for cotton and rayon.

5. *Ignition Time Tests*

Most flame spread and self-extinguishment tests specify one or two ignition times; others require forced ignition, i.e., exposure to the ignition source until the specimen burns. In addition, there are separate ignition time tests. Since the ignition time has to be established by trial and error, these tests tend to be time consuming. Two of these tests, which are part of the Australian[27] and British[33] standards, have been discussed in an earlier section.

Most ignition sources specified in flame spread and self-extinguishment tests provide ample energy to ignite fabrics. The activation energy of cotton has been calculated to be about 0.1 kcal/g,[56] while the luminous, premixed propane flame used in the German tests deliver about ten times as much energy. Another paper[36] states that in presence of a flame, untreated cotton is ignited by a heat flux of 0.35, FR cotton by 0.7, and wool by 0.8 cal/cm^2 sec. Without flame, the comparable heat flux values are 0.6, 2.0, and 2.0 cal/cm^2 sec, respectively. Theoretical analysis and modeling of radiative and convective ignition phenomena are the subjects of several studies.[62–64]

The parameters affecting ignition time with a given flame are specimen preparation, especially raising of pile fibers, preconditioning, specimen angle, point of ignition, fabric weight, and ability of the specimen, particularly thermoplastics, to shrink away from the flame. The effects of some of these parameters have been discussed in earlier sections.

Canada's ease of ignition test[65] specifies a horizontal, framed, 180 × 180-mm specimen, conditioned in an oven at 105°C. The 30-mm microburner flame is applied to the underside of the specimen, and the time to ignite determined by increasing the specimen exposure by 0.5-sec increments.

The German ease of ignition standard[66] also requires a horizontal specimen, 250 × 250 mm, mounted in a frame. The flame is the standard German 40-mm premixed propane flame and ignition is at the fabric edge. The frame is constructed so that the specimen can be ignited in three places (it is not clear how this affects the moisture content of the specimen which is conditioned at 20°C and 65% relative humidity). Ignition time or melting behavior is recorded.

The ASTM ease of ignition test, with the specimens oriented at 62½°, has been discussed earlier.[46] Ignition time of vertical, free-hanging, cylindrical specimens is part of the MAFT test,[10,67] which will be discussed in detail under heat evolution tests, below.

Small flames heat and shrink only small areas of thermoplastic fabrics, and cause less evasion of the flame and ablation than large flames, which tend to promote hole formation. Paper tabs attached to thermoplastic fabrics tend to be effective ignition sources because they move with the shrinking specimen.[5] There is some doubt whether they realistically represent conditions in real-life fires.

The ignition time of cotton and polyester–cotton fabrics was found to be essentially linearly related to fabric weight,[68] except for pile fabrics which ignited more rapidly than their weight would indicate. Ignition times of fabrics with low add-ons of FR finishes were found to be shorter than those of the untreated fabrics.[56] However, for FR concentrations close to those required to pass vertical tests, ignition time was found to be increased over that expected on basis of the weight increases,[69] due to the finish.

6. Heat Evolution Measurements

The criterion for most flammability tests for non-self-extinguishing fabrics is the flame spread rate. It is relatively easily measured. It was also assumed that the size of a burn injury in an apparel fire accident would be roughly proportional to the flame spread rate. However, flame spread rate measurements on fabrics retrieved from garment fires, measured on the CS 191-53 apparatus with forced ignition, showed no correlation with the size of the burn injuries of the victims.[37] Among the reasons for this perhaps unexpected finding may be that even though flames spread more slowly over heavy than light fabrics, they also produce more heat per unit area. Heat can rise inside garments and cause burn injury at some distance above the flame. In addition, heavy fabrics are generally more difficult to extinguish, as will be discussed in detail in a later section.

Heat transfer from burning fabrics has been measured by apparatus ranging from simple cylinders[70] to very sophisticated mannequins. However, we know only of two cases where heat evolution has been considered for regulatory purposes. One is the Australian standard AS 1176[27] in which heat developed by the burning specimen is measured by means of a copper rod located near the top of the cabinet. The other is the proposed general apparel flammability standard developed by the National Bureau of Standards for CPSC.[10,67] (CPSC has not decided whether to accept this draft standard, with or without modifications.) Because of the shape of the apparatus, it has been called the Mushroom Apparel Flammability Tester (MAFT).

The specimen is a cylinder, 178 mm in diameter and 300 mm high, simulating a pants leg or small skirt. It surrounds a copper cylinder, and is closed on top by a plate. The cylinder and top plate contain a total of 20 thermocouples which are used to measure the maximum rate at which heat is transferred to the inside of the burning specimen. Ignition is by means of a horizontal, 22-mm methane flame about 100 mm above the bottom edge of the specimen, for 3 or 12 sec. The reasons for the choices of this arrangement are explained in a backup report.[10]

Fabrics with a maximum heat transfer rate below 0.40 cal/cm^2 sec are in Class 1 and could be used in all garments. This class includes more fabrics than presently pass vertical self-extinguishment tests, e.g., the children's sleepwear tests. Almost all nylon and polyester fabrics are in this class, as well as fabrics made from modacrylic and matrix fibers and some of their blends with other fibers, and FR cotton and wool fabrics treated to levels somewhat lower than those needed to pass the children's sleepwear tests.(71) It has been shown by mannequin burns and other accident simulation experiments that such fabrics, while not necessarily self-extinguishing, have little injury potential.(10,72)

Fabrics which exceed this heat transfer rate could be used in normally fitting but not loose garments. An exception is fabrics which ignite in 1/2 sec or less, mostly cellulose-containing fabrics weighing less than about 100 g/m^2, which would be eliminated from the market. A large number of apparel fabrics have been classified by this method, and the potential effect of the standard on cost and fabric choice evaluated.(73)

The MAFT method was intended for pass–fail use but has been used to measure heat evolution from many FR and untreated fabrics. Several studies showed that the amount of FR finish on cotton fabrics to produce less than 0.40 J/cm^2 sec is 20%–40% less than that needed to pass the children's sleepwear test.(69,71,74) Marginally treated cotton fabrics can produce a narrow char but still a high heat transfer rate as measured on the MAFT. Agreement between the results obtained by two operators, and two MAFT apparatus, was found to be good.(75) Class 1 fabrics in combination with other fabrics can produce high heat transfer rates.(76)

Two other methods of measuring heat evolution were developed at the Textile Research Institute. In one,(77) the specimens are suspended in a wire cage inside an insulated cylinder, and the temperature rise on a thermocouple located in a dog-leg-shaped chimney is recorded. In this manner, gas temperature but not radiant heat is measured. Increasing concentrations of diammonium phosphate on cotton first increased then decreased the rate of emission but decreased the total heat in these experiments. However, both rate of heat emission and total heat first increased then decreased with increasing thiourea concentration on nylon fabric.

The TRI Airflow Calorimeter(78) measures heat evolution from burning fabrics indirectly by measuring the airflow through the burning chamber. The chamber is an approximately 300-mm-diameter cylinder, with the specimen suspended near the bottom. Air to the vertical chamber is supplied through a smaller, horizontal duct at its bottom. An anenometer inside this supply duct supplies airflow data which can be related to heat evolution by calibration with, e.g., methane or propane flames. One advantage of the method is that the heat measurement is taken in a zone which is never contaminated by pyrolysis products or molten debris. Studies with this instrument have shown that the total heat decreases with wool content in polyester–wool blends,(79) while the maximum rate of heat emission first increased then remained fairly steady with increasing wool content. Another paper considers methods of relating the time–heat curves to potential hazards[80] by combining the maximum heat emission rate, the time to reach this rate, and the time until the heat emission rate decreases to half of the maximum rate.

The major use of heat evolution measurements for determining fabric burn hazard has been in mannequin burns or similar simulations of fire accidents. Sensor systems on mannequins rank from papers which turn black at certain temperatures to sophisticated sensor systems which indicate the depth of burn.[64,81–85] The evaluation of the results is in terms of the area of the mannequin which would have received second-degree or deeper burns. The major findings of the referenced and similar studies are given below.

Garments move during burns, even on stationary mannequins, and this can affect the results. Many fabrics will extinguish when they make contact with the mannequin surface, or for that matter, with the skin of anaesthetized, shaved rats. Garments containing thermoplastic fibers shrink toward the mannequin. This may cause extinguishment, if they touch the mannequin, or more heat transfer if they just come closer to it without touching.

Looser garments burn faster and more completely. Belts act as temporary firestops and concentrate the heat toward the lower body.

Polyester and nylon garments, some acetate knits and wool fabrics, and modacrylics-blend knits exhibited low injury potential as single-layer garments. In combination with other fabrics, e.g., dress–slip combinations, nylon and polyester fabrics can cause more severe injury. The degree of injury depends on the contact between the thermoplastic and other layer during the fire. With good contact between a thermoplastic and cellulosic fabric, the burn injury pattern is similar to that of a cellulosic fabric with the weight of the two fabrics combined. If the outer garment burns, thermoplastic underwear which is at some distance from it may not burn at all, and may either offer some protection against the heat, or shrink to the mannequin and stick to it.

The most sophisticated of the mannequins which have been used for such experiments is "Thermoman."[84,85] The protoype is used by the U.S. Armed Forces to measure the protective characteristics of uniforms while the mannequin is pulled through a jet fuel fire. For civilian garment burns, the results are usually reported in form of B-80 values, the heat transfer patterns 80 sec after ignition in terms of areas which would probably receive second degree or deeper burns, with increasing weighting for burn depth. Within the framework of a comprehensive study of the burning behavior of over 60 apparel fabrics sponsored by the American Textile Manufacturers Institute,[84] over 300 male and (with properly instrumented attachments to Thermoman) female garments were burned on Thermoman. The results were compared with those of 15 other laboratory test methods. These include several heat evolution tests: the TRI air flow calorimeter, MAFT, and tests employing single and banks of heat sensors; and char length, char area, and weight loss and flame spread rate measurements. Correlation coefficients for B-80 values (regardless of garment type) and laboratory test results did not exceed 0.70. One reason for this lack of correlation is that no single fabric was used in all garments, but garments were made from the appropriate fabrics, e.g., blouses only from blouse fabrics, and slacks only from the heavier slack fabrics. As expected, the B-80 values for the looser garments and those which covered both the upper and lower body such as robes and dresses, were higher than those obtained with skirts or blouses which covered only part of the body, or graments which fitted tightly. Similar results were found in the analysis of real-life accidents.[10,37]

Because mannequins do not move but burn victims usually do, an attempt was made to compare heat transfer from burning fabrics when they hang freely and when they make contact with a sensored surface during the burn.[72,86] The apparel fire modeling apparatus (AFMA) was constructed for this purpose. It consists of a semicylinder, 600 mm high and 180 mm in diameter, and is almost completely covered with heat sensors. Semicylindrical specimens are held loosely near the AFMA surface, and either allowed to burn without interference, or brought in contact with the AFMA surface during a predetermined stage of the fire, to simulate the movement of a victim which causes garment-body contact. Many fabrics extinguish after this movement but the heat delivered to the sensors varies widely. Fabrics which continue to burn are those which make poor contact, e.g., terry towels, and those which shrivel away from the AFMA surface, such as acetate and acrylics. Fabrics ranked roughly as follows in order of increasing heat transfer to the AFMA: nylon, polyester, wool, acetate knits, cotton and polyester–cotton blends, woven acetate, polyester–cotton terry cloth, and acrylic fabrics.

Some investigators use time lapse photography and other measurements on mannequin burns to estimate their severity, rather than heat transfer measurements. A "desirability value" was based on measurement of maximum flame area, maximum flame area growth, total area destroyed, and total burn time of dresses on a mannequin, and of the same measures plus weight loss and maximum rate of weight loss for flat specimens.[87,88] For a number of modacrylic–acrylic blends, desirability values correlated with char length measured by the children's sleepwear test but not with oxygen index (OI) values.

One investigator found little difference in heat transfer from fabrics contaminated with flammable liquids and made from either cotton or polyester–cotton blends, though in the uncontaminated fabrics, the heat transfer decreased slightly with polyester content.[89] However, mannequin burns[90] and AFMA experiments[91] have shown that less heat is transferred from self-extinguishing fabrics contaminated with flammable liquids than from flammable fabrics. The area wetted by a given amount of gasoline was, in the first approximation, inversely related to fabric weight, as was the evaporation rate.

7. *Extinguishability*

An analysis of human behavior in apparel fire accidents[92] showed that the most prevalent first reaction was running, followed by beating the flames with the hands, removal of clothing, rolling on the ground, wrapping the victim with a rug, and dousing with water. Often more than one of the above extinguishing methods were used. Removing the clothing was apparently most effective in keeping the injury small, while the often recommended wrapping in a rug or similar heavy fabric resulted in larger injuries, perhaps because the heat is confined near the body. The data did not permit analysis of the effect of fabric and garment variable on the success of the attempts to extinguish the fire.

Extinguishability of fabrics can be considered a function of oxygen exclusion and contact with a heat sink. This contact can be on one side of the fabric, when the garment comes close to the body of the victim due to fit, shrinkage, or movement of the victim. Many fabrics stop burning upon contact with a mannequin or living skin.[72,82,86] Contact with both sides occurs in slapping, rolling, etc. In some cases, rapid movement may extinguish the flames; in others, it may enhance them. The mechanistic and material aspects of the extinguishment of burning textiles are considered in the first of a series of papers from the University of Maryland.[93] This series also contains reports on experimental work in this area.[94–96]

No tests for fabric extinguishability have been considered by standard developing organizations, but in recent years, a few laboratory studies have been published. When attempts were made to extinguish burning garments on a mannequin by slapping with gloved hands, lightweight cellulosic fabrics could be extinguished easily even though they burned at a rapid rate.[82] However, heavier fabrics required that the hand remain in place for a distinct time, perhaps more than a second. With an ungloved hand, this would probably be too painful. Fabrics containing polyester–cotton blends stuck to the glove when it was withdrawn, and sometimes burned more intensely because of this rapid movement. The relationship between fabric weight and time of two-sided contact between heat sink and the burning fabric to achieve extinguishment was confirmed by more controlled laboratory experiments.[97]

No such dependence on weight was found in the already discussed extinguishment experiments in which only one side of the burning fabric contacted a heat sink, the AFMA surface.[86] In these experiments the contact was made when a heat sensor on the AFMA surface registered a temperature increase of 5° C. At this point, light fabrics were burning at a fast rate, and heavier fabrics were burning slowly but produced more heat.

The TRI flammability analyzer[60,94,98] permits measurement of a steady flame spread rate at various oxygen concentrations, using a specimen mounted on circular tracks. A steady flame position at 3 o'clock (achieved by turning the track at the rate of flame progress) can be considered an upward burning flame; at the 9 o'clock position, the flame burns downward. Oxygen concentration can be plotted against burning rate at any one position. An extrapolation of the resulting linear relationships to zero burning rate results in a value called the "intrinsic OI." This value is assumed to be related to the sensitivity of various fabrics to extinguishment by reducing the oxygen concentration. The TRI fabric flammability analyzer was also used to study the effect of the presence of a heat sink near the specimen on the burn rate–oxygen concentration relationship.[95] Linear burning rate decreased with decreasing distance from the heat sink but raising the temperature of the heat sink from 37 to 110° C had no significant effect.

The air patterns behind a running person were investigated and heat evolution from burning fabrics exposed to air streams of varying velocities was measured.[96] Behind a running person, the air is still and flames can burn vigorously even though they extinguish in front. Some fabrics showed first increasing, then decreasing heat evolution with increasing air speed, for others heat evolution decreased fairly linearly with air speed. Fabrics which had low heat evolution in still air were extinguished by relatively slow air streams. Air streams did not extinguish fabrics which form chars and which had started burning on the inside. Cellulose fabrics which has been treated with FR finish to levels where the flame path is narrow were easily extinguished by moving them.[99] This may indicate that they may not represent a major hazard even though they may not pass the more severe flammability tests.

8. *Effect of Laundering, Soiling, and Weathering on Flammability*

The children's sleepwear tests[2] require testing of the fabric in its original state or after one laundering and tumble drying cycle, as well as after 50 cycles. They also require labeling of the garments with laundering instructions, including recommendations of detergent type, and warning against the use of soap, bleaches, softeners, drying in sunlight, etc, where indicated by the nature of the FR finish used or the fiber characteristics. These requirements lead to a large number of studies, particulary on the effect of laundering conditions.

Increase in char length of laundered fabric can be due to loss or chemical changes of the FR finish, or to build-up of calcium and magnesium salts. Salt build-up occurs especially when soap or low phosphate detergents are used, especially with hard water. When garments are worn between launderings, much less build-up occurs.[100] The standards may thus be more severe than real-life conditions because they specify 50 laundering cycles without wear. On the other hand, the standards are conservative by specifying a high phosphate detergent and relatively high water temperatures; both factors result in less salt deposit than low phosphate detergent and low temperatures in the washing machine.

The many studies carried out on the effect of laundering can be summarized as follows.

FR cotton fabrics were found more sensitive to laundering conditions than other FR fabrics.[101] Nylon, modacrylic, and matrix fabrics passed the children's sleepwear standards after repeated laundering with carbonate detergent and soap, even when the water was hard, while FR acetate and especially FR cotton showed increased char length. A recent paper showed that repeated laundering and wear of commercial nylon, polyester, and FR 80/20 acetate–nylon nightgowns did not affect the char length.[102]

The calcium carbonate concentration at which FR cotton fabrics failed the children's sleepwear tests depended on the manner in which the salts were deposited, the FR treatment type, and a variety of laundering conditions. The upper level for passing seemed to be 2% of calcium, based on the weight of the fabric.[103] Under similar application conditions, fabrics treated with phosphonium-based FR finishes contained more calcium than THPOH-ammonia-treated fabrics.[104]

Both magnesium and calcium deposits can be removed by the use of high phosphate detergents, water softeners, or acid rinses.[105,106] Many housewives would resort to such steps because the deposits would give the garments a gritty feel and discolor them. However, frequent use of acid rinses can degrade some FR finishes.

Other factors which affected the char length of FR-treated cotton fabrics were the sequence of placing clothes and detergents into the washing machine, water temperature, drying procedures, etc.[102,104–106] Bleaching and fabric softeners also can increase char length of FR cotton fabrics.[107–109]

Various workers studied the interaction of soil and FR treatments. Some of their findings follow.

Milk and urine had little effect.[110]

Baby oil and talcum powder, in rather large concentrations, increased flammability.[111]

Baby oil and silicone lotion caused FR cotton to fail the children's sleepwear test at about 7.5% add-on, and nylon at 15%.[112]

FR finishes were found to increase soil pickup slightly, but soil removal depended on finish type.[113] The soils applied did not affect the char length or OI. Similar effects were found with mildew stains and mildew stain removal.[114,115] Weathering seemed to affect some finishes but others were not affected by carbon arc exposure.[116]

A series of papers described the effect of various hospital laundering procedures on FR cotton, polyester–cotton, and polyester fabrics.[117–120] Satisfactory procedures for laundering and sterilization could be worked out; among several FR finishes for cotton, THPOH-ammonia performed best.

Drycleaning did not affect the char length of FR fabrics such as cotton, wool, Nomex, and modacrylic.[121]

In the above studies, the evaluation of the specimens consisted primarily of measuring the char length. Several investigators also measured OI values and performed DSC, TGA, and DTA analyses.[115,116] No changes in autoignition time and DSC and TGA results were found due to presence of calcium stearate on cotton fabrics.[122] However, the relationship of mass transformation rate and oxygen concentration on the TRI fabric flammability analyzer was changed by the presence of this salt on FR cotton, PFR® rayon, acrylic, and polyester–cotton fabrics.

9. *OI and Other Research Methods*

The concept of the oxygen index was introduced for plastics[123] but it was quickly adapted to textiles as well. The lack of rigidity of textile specimens presented some experimental difficulties, and required use of sample holders.[124] Investigations of the effect of chimney length,[125] and direction of gas flow, sample size and orientation, and sample holder have been carried out[126] and indicate the need of close control of all factors to obtain comparable results. Increasing cotton fabric weight and tightness of construction resulted in increasing OI values, as did moisture content above 6%. FR cotton with an OI of 28 generally passed the children's sleepwear tests, while those with OI of 26 or less did not. In the 26–28 range, passing or failing depended on such factors as ignition time. For other fibers, a poor correlation between OI and char length was found.[127] The OI test was found to be more useful than char length for the characterization of high-temperature resistance of fabrics for protective clothing.[128] Fabrics with an OI of less than 25 burn in air, those with OI of 25–30 burn in hot air and drafts, and those with OI exceeding 30 can be considered nonflammable under conditions normally encountered in situations requiring some heat protection.

A statistical analysis of OI results showed that determination of the point at which combustion is barely supported, the usual criterion in OI tests, was the major cause of variability between operators.[129] Modified versions of the OI apparatus were used to study the sewing thread[130] and carpets.[131,132]

The results of vertical and 45° tests, of ignition in a muffle furnace experiments, and of OI, DTA, TGA, and smoke obscuration tests on untreated and FR-treated cotton and polyester–cotton fabrics, and on polyester, wool, and Nomex fabrics, were compared.[124,133] Cotton fabrics discolored and lost strength at 300° C in the muffle furnace. At 400° C, untreated and FR-treated samples ignited. Ignition time decreased with add-on of FR finish. Nomex did not ignite, discolor, or lose strength at 400° C. The decomposition temperatures were lowered by softeners, dyes, and FR finishes, as was the lowest point of the endotherms. OI results were also sensitive to dyes and finishes. Fabrics which contained polyester produced more smoke than cotton fabrics.

Several other authors found increases in OI due to dyeing with acid, reactive and direct dyes, as well as flourescent brightening agents.[134–138] A variety of dyes lowered the decomposition temperature of cotton, by as much as 50° C. Reactive dyes decreased ignition time but had no effect on MAFT heat transfer rate results.

A polyester fabric was dyed with 19 disperse dyes[137] and tested with the children's sleepwear tester, the MAFT, and the TRI flow calorimeter. No dye had a significant effect on all parameters measured. The majority of dyes increased ignition time. Disperse azo dyes seemed to have more adverse effects than quinophtalone or anthraquinone dyes.

OI results were found to be considerably more sensitive to concentration of FR finishes on fabrics containing cotton, as well as have better reproducibility and precision than other flammability tests.[138] The char length and OI of a variety of blends of Nomex, modacrylic, and PVC fibers with cotton, wool, rayon, and polyester were measured.[139] The behavior of the blend is not readily predictable from its fiber content. As expected, the amount of FR finish a Nomex–cotton blend needed to obtain self-extinguishment increased with increasing cotton content.

A fair correlation between OI and MAFT heat transfer rate was found for cotton fabrics treated with three FR finishes at levels below those needed to pass the children's sleepwear test, and dyed with five dyes.[140] Other experimentally treated fabrics with OIs of about 26 did not ignite in the MAFT test.[127]

10. Evaluation Methods for Specific End-Use Items

10.1. Blankets

In June 1970, the U.S. Department of Commerce published a notice in the Federal register that a flammability standard for blankets may be needed.[141] The National Bureau of Standards developed a draft standard,[1,8] but before it was published, the industry decided to voluntarily comply. No further action was taken by the Department.

The proposed test method uses the CS 191-53 cabinet and burner. Specimens, 70 × 70 mm, are oven dried at 105° C. A well-defined paper with a 51-mm circular cutout is centered over the specimen, and covered with an aluminum plate with 64-mm-diameter cutout. The assembly is placed on a horizontal stand and the flame allowed to impinge for 1 sec. If the paper collar is charred or burned, the specimen fails. Most blankets except those containing a large percentage of cellulosic fibers in pile constructions pass this test.

10.2. Carpets

Flammability evaluation methods for carpets can be divided into two classes: methods which determine the resistance of carpets to small ignition sources, such as a dropped match or a spark from a fireplace, and methods which evaluate the performance of carpets in building fires.

There are two U.S. standards for surface flammability of carpets and rugs, FF 1-70[142] and FF 2-70,[143] and a Canadian standard similar to FF-1.[144] FF-1 applies to all carpets which are larger than 2.23 m^2 and have one dimension exceeding 1.83 m. FF-2 requires that carpets smaller than that can be sold even though they do not pass the FF-1 test but must be labeled accordingly.

Specimens 229 × 229 mm are dried in an oven at 105° C and cooled. They are placed in an enclosure which is open on top. A steel flattening plate, with a 203-mm-diameter cutout, is placed on the specimen. A methanamine pill (No. 1588, Ely Lilly and Co.) is placed in the center of the specimen and ignited. This pill burns up to 120 sec, and the heat flux varies with time but a flux of 0.4 W/cm^2 was measured for about 30 sec.[145] The carpet passes if in seven of eight specimens the char does not extend to within 25 mm of the flattening plate. The standard eliminated approximately 10%–15% of the carpets on the U.S. market at the time of introduction. There were mostly low-priced carpets with open constructions which generally have low durability.

The British standard employs a similar concept.[146] Specimens are conditioned at 20° C and 65% relative humidity. A hexagonal nut is heated to 900° C and placed on the specimen for 30 sec. The radius of the charred area is measured on both sides of the specimen. Chars of 35-mm radius are classified as "low radius of effects of ignition," 40–70 mm as medium effect and 70 mm or more as high effect.[147]

The West German method[148] specifies a vertical specimen in contact with an asbestos cement board. The specimens measure 345 × 104 mm and are conditioned at 20° C and 65% relative humidity. Ignition is with a 40-mm premixed propane flame on the surface of the specimen, for 5 or 15 sec. A separate standard[149] divides the carpets into three classes, according to ignition time, afterflame time, and whether a trip thread 250 mm above the point of ignition is burned through.

In general, tightly constructed, low-pile carpets can pass these tests. Carpet backing and pads affect the results but the above standards do not require testing of carpet and pad as a unit, since this would place a heavy burden on the seller of the carpets. Wool ranks relatively high in these tests; acrylics can be upgraded by addition of PVC. Cellulosic carpets, which had only minor use in the U.S. when the standard was introduced, ignite relatively easily. A study with nylon carpets using the FF-1 method[150] showed that fewer failures occurred with jute than with polypropylene primary backing, and with jute secondary backing than with urethane latex. Alumina trihydrate incorporated into the adhesive latex reduced the number of failures.

Horizontally mounted shag carpet specimens were ignited with methanamine pills in an OI test apparatus.[131] Carpets with OI of 18–20 failed FF-1, those with OI of 22 regularly passed it. Dye bath residues, especially silicones, reduced the OI. The advantage of the method was that it requires only small specimens, and that the effect of backing could be minimized. It thus appears applicable to research experiments.

In another study, carpets were ignited in the OI apparatus with methanamine pills and the rate of flame front movement measured.[132] The "limiting OI" was the oxygen concentration at which the carpet burned more than 25 mm. The authors state that it is difficult to establish meaningful pass–fail criteria with this method.

Carpets were mounted on the TRI fabric flammability analyzer[151] and the relationship between oxygen concentration and burn rate established. This relationship was similar when the ignition sources were the British heated nut or two or three methanamine pills, but more oxygen was required for a given rate when only one pill was used. Carpets used in these experiments contained viscose, acrylic, modacrylic, nylon, propylene, polyester, and wool, in plush, loop, and shag construction and varying in pile height and backing constructions. OI results varied from 16.7 to 25.7. The viscose carpet had the lowest OI. Acrylic and nylon carpet OI results varied according to pile density, while the results for the polypropylene carpets were not affected by construction. The wool and modacrylic carpets had the highest OI results. Smoke density was found to be a function of fiber content and backing material.

Chairs containing urethane foam and cotton batting were burned in a room with noncombustible walls and ceilings, with carpets which passed the FF-1 test.[145] The chairs burned vigorously, but the fire did not spread beyond the carpet area adjacent to the chairs. However, carpets which pass FF-1 have been found to spread fire under more severe fire conditions. Such a carpet was used in the corridor of the Baptist Tower in Atlanta in which a fire spread from a burning room along the corridor. In such cases, heat is reflected from the ceiling or smoke layers near the ceiling and upper portions of the wall, and heats the floor covering. The heated carpet can burn more vigorously than carpet at normal temperatures, and in severe cases, flashover can occur. Several test methods simulate the scenario to varying degrees.

The newest of these methods is the flooring radiant panel test (FRPT).[152] The specimen, 1070 × 250 mm, consists of both the carpet and the pad or floor on which it is to be used. It is conditioned at 23° C and 50% relative humidity and mounted in a horizontal frame. Above one end of the frame is a gas-fired, radiant panel, inclined 30° to the horizontal. This results in a heat flux on the specimen varying from 1.1 to 0.1 W/cm^2. After exposure to this flux for 2 min, a 50-mm propane pilot burner flame is brought in contact with the specimen at the high flux end for 10 min. The total distance burned, i.e., the point of farthest advance of the flame in this environment of decreasing flux, is measured and called the "critical radiant flux." For corridors and exitways in hospitals and nursing homes a minimum critical radiant flux of 0.45 W/cm^2 is recommended; for corridors and exitways in other occupancies except one- or two-family houses, 0.22 W/cm^2.[153] The critical radiant flux of carpets which were involved in corridor fires, e.g., the Baptist Tower fire, was about 0.1 W/cm^2. In the previously mentioned experiments with burning chairs and wood cribs,[145] the critical radiant flux of carpets was found to correlate with the extent of carpet burning.

The effect of carpet and pad interaction was studied.[154] Fires were started in a room which was open to a corridor. No flashover occurred with the carpet alone but flashover did occur when the carpet was mounted over a pad. The nylon carpet alone had a critical radiant heat flux of 0.24 W/cm^2, the carpet over various pads from 0.120 to 0.126 W/cm^2. The pad experiments produced more smoke than the carpet alone during the smoldering stage but not during flaming.

A mathematical model for the FRPT has been worked out[155] and critical radiant heat flux values found experimentally compared with heat flux values calculated for large fires in corridors. It appears that the critical radiant heat flux value is a valid measure of fire hazard if the fire advances against the flow of air.

The Steiner Tunnel test is widely used for building materials as well as floor coverings.[156] The specimen measures 7.3 × 0.5 m and is mounted, face down, on top of an insulated, horizontal tunnel. Two upward-pointing gas burners are at one end, and an exhaust blower on the other. Test duration is 10 min. The criteria are flame spread rate over the last 5.9 m of the specimen, smoke obscuration, and fuel contribution. Classification is based on 0 for asbestos cement board and 100 for red oak flooring. Results of an interlaboratory test for four carpets show better reproducibility for flame spread ratings than for smoke development and fuel contribution.[157] Some of the variability may have been due to differences in the construction of the tunnels. Canada has adopted this test for carpets, but with the specimens mounted on the floor.[158] This prevents melting and ablation of thermoplastic carpets mounted face down.

A third method was developed by Underwriters' Laboratory.[159] Specimens measure 610 × 96 mm and are conditioned at 21°C and 50% relative humidity. A gas burner at the fire end of the chamber is directed against the horizontal specimen and is applied for 12 min. Flame front propagation rate and afterflame time are recorded.

The Federal Aviation Administration applies its standard for interior cabin materials[17] to carpets. It is based on the vertical, 12-sec edge ignition method 5903.

10.3. *Curtains and Draperies*

Curtains and draperies in public occupancies are usually required to be self-extinguishing. Local or state fire authorities determine the method to be used; many choose NFPA 701.[20] Other local authorities have designed their own tests which should pass the same fabrics in most cases. Many tests described earlier are used for both apparel and curtains.

Thirty-eight full-scale curtain fire experiments were conducted in a instrumented room.[38] Samples of common wall cover and ceiling materials were mounted near the curtains. Rate of flame spread, air temperature, radiant heat flux, and smoke and toxic product evolution were measured. Curtains with rapid flame spread often did not ignite the wall and ceiling materials, while heavy curtains which burned slowly ignited many or all of them. Heat evolution seems thus to be more important than flame spread rate when secondary ignition is considered.

10.4. *Mattresses*

A majority of fires starting in mattresses are caused by dropped cigarettes.[160] The cigarette may smolder slowly through bedding, mattress cover, and may continue into the filling material. The fire may smolder for hours, and the common mode of injury is smoke inhalation rather than burns. Recent reviews are available on smoldering combustion[161] and on mattress flammability.[162,163]

The mattress flammability standard, FF 4-72,[3] covers all mattresses sold in the U.S. Full-size mattresses are used in this test. Conditioning is at more than 18°C and less than 55% relative humidity. Burning cigarettes (king size, nonfilter) are placed at various places on the mattress, such as welt edge, and tufted and flat areas. Cigarettes are placed on the bare mattress, and on the mattress between two sheets. Char length exceeding 51 mm disqualifies the mattress.

Canada's mattress test does not require testing of the whole mattress.[4] The specimen is a 300 × 300-mm piece of a mattress, which is conditioned at 21°C and 35%–50% relative humidity. The specimen is placed in a tightly fitting box, 300 mm high. An indentation tester is applied so as to produce a 1.30–1.35-mm indentation with a force of 2.45 N/cm^2. A burning cigarette is placed near a stitching line and allowed to burn out. Specimens pass if the char length does not exceed 50 mm in any direction for the cigarette location, and if there is no active combustion 10 min after the cigarette burned out.

Thermoplastic backcoatings decrease the heat transfer from the burning cigarette through the mattress ticking to the filling.[164] Diammonium phosphate, various pigments, etc., are also effective, as is boric acid treatment of cotton batting.[165]

Mattresses have frequently been ignited on purpose with matches or sheets of newspaper in prisons and mental institutions. The rapid spread of smoke and toxic pyrolysis products in such confined quarters has caused several multivictim fires. The U.S. Bureau of Prisons specifies mattresses filled with FR-treated cotton batting (borax–boric acid treatments) which resists both smoldering and flame ignition. Ordinary urethane foam is moderately resistant to smoldering ignition by cigarettes but not to flame ignition.

In an invesigation of mattress fires,[166] a room was instrumented to measure heat evolution, smoke density, CO and CO_2 concentration, and oxygen depletion. Beds were made up with various combinations of sheets, blankets, mattress covers, and filling materials. A dummy was placed in the bed to assure natural configuration of the bedding materials.

Twelve of the 22 bedding combinations ignited from a single cigarette. In some cases, life-threatening conditions occurred in as little as 5 min, and in all but one case (when all materials were flame retardant) within 30 minutes. Except for this one case, zero light transmission occurred in these tests. Laundered sheets ignited from a cigarette, but new sheets did not.

A more recent report includes an extensive literature search on mattress fires and describes additional mattress fire experiments.[162] The ignition source was a small polyethylene waste basket filled with typical trash, placed against the edge of the mattresses, and ignited with one match. Textile items used in these experiments included sheets, and rayon, untreated and FR cotton, polyester, and PVC-coated nylon and cotton ticking, cotton–polyester and cotton–polyester–nylon pads. Mattress cores were polyurethane, butadiene–styrene latex, and neoprene foam.

In the flaming ignition mode used in these experiments, no effect of the ticking was found. FR cotton batting–innerspring combinations resulted in the lowest heat, smoke, and toxic gas development. Neoprene was considered acceptable in heat and toxic gas evolution but not in smoke development. Three of the four polyurethane mattresses performed poorly on all three scores, and one polyurethane and one latex mattress caused room flashover. Treated cotton batting[162] and neoprene innerliners[167] have been shown to reduce mattress flammability in full scale tests.

10.5. Upholstered Furniture

Fires in which an upholstered furniture item was the first item to ignite have been estimated to have caused 1500 deaths, 4500 civilian and 200 firefighter injuries, and $300 million direct property loss in 1978.[168] About 75% of such fires are started by dropped cigarettes, and many never go beyond the smoldering state.

Upholstered furniture items are a complex arrangement of a wide variety of materials. Cover fabrics are rayon, cotton, acetate, nylon, polyester, polypropylene, wool, silk, or blends thereof. Filling materials can be untreated or FR polyurethane or neoprene or latex foams, untreated or FR-treated cotton batting, polyester–cotton or 100% polyester batting, down or feathers, etc. Different filling materials are usually used in the seat, sides, and back, and on the inside and outside of the chair. Finally, there is a wooden frame, sometimes springs, etc.

CPSC is considering an upholstered furniture flammability (cigarette ignition) standard, PFF 6.[169] It specifies two tests. In one, a mock-up of a line of upholstered furniture is tested, with the materials in the mock-up arranged exactly as in the furniture line. In the other, cover fabrics are classified.

The mock-up consists of a 610 × 610-mm cushion (a common production size) and two vertical sides, 650 × 370 and 545 × 370 mm. The cover fabric, welt cord, and filling materials are arranged in the same manner as in the seat, back, and sides of the production furniture. Burning, king-size, nonfilter cigarettes are placed on all horizontal surfaces of the mock-up, including the top of the arms and back if they are wide enough, flat and tufted areas of the pillow, at the welt edge, and the crevices formed by the seat and the two side panels. If the char length from any point of the cigarette exceeds 76 mm, the item fails and one or more of the components must be changed.

Many furniture manufacturers use hundreds or even thousands of cover fabrics, over fairly standardized filling constructions. To eliminate the need for testing every fabric in mock-up form, the proposed standard has a provision for classifying fabrics. The fabric to be tested is attached to two glass fiber boards, 204 × 204 and 204 × 116 mm. The boards with the fabric attached to them are placed into a sample holder, the larger vertical, the smaller horizontal, so that a 204-mm crevice is formed. The cigarette is placed into the crevice and allowed to burn out. If the char length exceeds 76 mm, the fabric is Class D; if 38–76 mm, Class C. If it is less than 38 mm, the fabric is tested again, with the vertical glass fiber board covered with untreated cotton batting. If the char length now exceeds 38 mm, the fabric is in Class B; if not, in Class A.

The standard also defines a "Standard Class B" (a 125-g/m^2 cotton or polyester–cotton sheeting) and a "Standard Class C" (a 245-g/m^2 cotton ticking) fabric. If any Class A or the standard Class B or standard Class C fabrics pass the mock-up test with any given filling material arrangement, any Class A, B, or C fabric, respectively, can be used with the same filling material arrangement. However, because Class D fabrics present the highest smoldering ignition risk and cannot be represented by a standard fabric, all constructions using Class D fabric would have to be tested in mock-up form.

Class A fabrics are generally vinyl, wool, nylon, polyester, and polypropylene fabrics, in the heavier weight ranges. The thermoplastic fabrics dissipate the heat by melting. Class B fabrics are the lighter-weight thermoplastics, blends of thermoplastics and cotton, and some lightweight, tightly woven cellulosics. Class C and D include most high-cellulosic-fiber-content fabrics, with D fabrics in the upper-weight ranges.

Traditional FR treatments do not seem to improve the smoldering ignition resistance of cellulosic fabrics. However, vinyl–vinylidene backcoatings,(170) and boric acid and elemental sulfur,(171) have been reported to increase smoldering ignition resistance. Other means to upgrade smoldering ignition resistance are the use of polyester batting, and, for some particularly smolder-prone fabrics, use of aluminized surfaces between cover fabric and filling material, for better heat dissipation.

A backup report explains the choices of the provisions of the proposed standard(172) and reports results of interlaboratory evaluation of the mock-up and fabric classification tests. Two preliminary studies of the cost effectiveness of the standard have been published,(173,174) and they use different approaches and produce somewhat different results.

The Upholstered Furniture Action Council (UFAC), an upholstered furniture industry group, presented a counterproposal to CPSC in form of a voluntary standard to produce furniture with low propensity to ignite from cigarettes.(175) The Commission decided, in December 1979, to let the industry proceed with this standard for one year. UFAC proposes five test methods, for individual furniture components rather than assemblies of fabric and filling materials in mock-up form. Fabrics are divided into two classes by a test similar to the one of the proposed CPSC standard, but using polyurethane foam as a substrate rather than glass fiber board. Class 2 fabrics are medium- and heavy-weight cellulosics, Class 1 all other fabrics. The welt cord, filling, barrier, and decking materials would be tested in the same arrangement with standard fabrics. The proposal also specifies certain combinations of materials which can be used, e.g., untreated cotton batting can not be used in UFAC-labeled furniture, and polyurethane foam can not be used in cushions under Class 2 fabrics without a barrier material which passes the test. Untreated cellulosic welt cords do not pass the test, and could not be used.

A British standard[176] also requires mock-up testing but specifies both cigarette and flame ignition. The cover fabric specimens measure 1350 × 700 mm, the filling material specimens measure 500 × 500 × 75 mm. The cigarette is placed into the crevice, and the time to self-extinguishment of smoldering or flaming noted. Additional tests are conducted with a butane burner flame directed at the crevice, with increasing gas flow and ignition time. Finally, wood cribs with weights increasing from 9.25 to 132 g are placed in contact with the horizontal and vertical sides of the mock-up. No pass–fail criteria have been proposed as yet.

The state legislature of California required a state standard for upholstered furniture which covers cigarette as well as flame ignition, and is applicable to furniture used in homes and public occupancy.[177,178] Pieces of furniture are acquired from stores and tested for cigarette ignition, using the same cigarette as the proposed federal standard but additionally specifying the burn temperture at 545°C. All surfaces and crevices are tested. Flaming or a 51-mm char length present failure. At present, failing furniture must only be labeled. However, the intention is to insist on compliance when it becomes technologically practicable.

However, all furniture must pass the following tests in California.

The resilient filling material must pass Method 5903[15] before and after aging. Char length must not exceed 154 mm; afterflame time must be less than 5 sec, and afterglow time less than 15 sec. Shredded foam, feathers, etc. are encased in ticking for the test. Expanded polystyrene beads are placed into a metal basket and tested with methanamine pills; the weight loss must not exceed 5%. Cotton batting is tested vertically, without a frame, and polyester fiber batting is tested by CS 191-53[39] with a 5-sec ignition. Burn time must exceed 10 sec. Cover fabrics are also tested by CS 191-53, and the burn time must be more than 4 sec for smooth, and more than 7 sec for napped or pile fabrics. Finally, the filling material must withstand cigarette ignition, with a maximum char length of 50 mm.

The provisions of MVSS 302[50] are applied to automotive upholstery. The surface and filling materials are tested separately unless they are bonded or sewn together. Aircraft upholstery is tested under an FAA standard[17] which specifies use of Method 5903.[15] Char length must not exceed 202 mm, afterflame time 15 sec, and melt drip must not burn more than 5 sec after falling.

A recent review paper on furniture flammability mentions 68 references.[179] Much of the work was performed by or under sponsorship of the California Bureau of Home Furnishings. In one of their studies, they tested the cigarette ignition resistance of 88 upholstery fabrics, each mounted over 12 filling materials.[180] The crevice arrangement of the proposed CPSC standard[169] was used. Ignition was defined as 51-mm char length or obvious smoldering. The interaction between cover fabrics and filling materials was clearly demonstrated. Fabrics with high thermoplastic fiber content, and wool and silk fabrics, ignited over few filling materials, while heavy cellulosic fabrics ignited with all filling materials. Untreated cotton bating ignited with about 80% of the cover fabrics, FR cotton batting with only about 50%, polyester fiberfill with 10%–18%, depending on fiberfill type, and untreated polyurethane foam with about 60%. No systematic effect of cover fabric weave or construction was established. The CPSC fabric classification test placed 17 of the 88 fabrics into Class A, 13 into Class B, and 58 into Class C or D.[181] When the CPSC standard glass board was replaced with Pyrex fiber filter pads, more fabrics passed, indicating perhaps an effect of the binder resin in the glass fiber board.

The flash fire propensity[182] and the relative toxicity of pyrolysis products[183] of upholstery fabrics were studied, as well as the ignition characteristics of some home furnishing and aircraft interior fabrics under radiant heat.[184,185] However, these experiments were carried out on the fabrics only, not in conjunction with the filling materials. Another investigation studied upholstery fabrics in combination with foam, and found that burning behavior and CO emission could not be predicted from the performance of the individual components.[186] Furniture cover fabrics were tested over flat, horizontal foam pieces. None of the chosen fabrics ignited from cigarettes in this configuration. Wool and vinyl did not ignite from matches or methanamine pills. All other fabrics—rayon, cotton, acrylics, acetates, and blends thereof—ignited with matches when bone dry. Fewer fabrics ignited at 65% relative humidity with matches, but all fabrics ignited with methanamine pills. Smoke emission was least from two acrylic, one acrylic–rayon, and one wool fabric and most from vinyl.

The major factor in flaming ignition of upholstered furniture is the filling material.[187] Cover fabrics which char, such as cotton and wool, can delay flame or radiative ignition of the filling material, while thermoplastic fabrics can melt away and permit rapid access to the filling. Thus thermoplastic fabrics, which generally offer better resistance to cigarette ignition, the most frequent accident scenario, have less resistance to flame ignition.

Chairs which were identical except for the cover fabrics were ignited with folded newspapers placed in the seat.[188] The tenability limits, i.e., the time to incipient incapacitation of victims due to heat, CO_2 and CO concentration and oxygen depletion, and/or smoke obscuration, were calculated from the measurements taken in the test room. Tenability limits were exceeded at 150 and 215 sec after ignition for chairs with polypropylene and nylon fabrics over a layer of untreated cotton batting and a polyurethane core. With the same filling arrangement but a cotton cover fabric, the limits were not reached but temperatures in the room were higher, for a longer time. Chairs in which the foam core was not wrapped with cotton batting performed worst on all scores.

Only one study appears to be available which considers the effect of use and aging of furniture on flammability.[189] The char length caused by cigarettes placed on chairs for 3 and for 10 min was measured. It was often but not always slightly larger after two years of use in college facilities than on the new chairs.

10.6. Protective Clothing

Many fabrics have been developed for protection against heat, ranging from the heat flux levels to which some industrial workers are exposed to suits for race drivers and military pilots and firefighters who may be exposed in fuel fires. The protective garment must not burn, it must not lose its integrity by melting or forming of brittle chars, it must retain its shape, so it must not shrink or soften, and it must insulate against the heat. All this should be attained with minimum physiological load on the wearer.

Fabrics which pass the common, vertical self-extinguishment tests are considered satisfactory for the lower-level exposures, e.g., firefighters turn-out coats and industrial workers. Method 5903[15] is frequently used for these materials. For higher heat exposures, the LOI can be used as a screening tool.[128] Wool, Nomex, Kevlar, Kynol, Fypro, FR cotton and rayon, etc. have OIs of 25–30. Such fabrics will burn in hot air, with a good draft. Fibers like Durette, polybenzidazole, polyimide, PBI, and specialty aramids have OIs exceeding 35.

Several test methods have been developed to measure the insulative properties of the fabrics, as well as the retention of their integrity. Heat sources range from radiant panels to gas burners or a combination of the two, depending on the type of fire to be simulated. Heat sensors range from thermocouples to various heat sensors, to a simulated skin consisting of a mixture of alpha cellulose, urea formaldehyde, and silica powder which has the optical and thermal characteristics of real skin.[190] In general, insulative protection increases with fabric thickness for fabrics with a reasonably dense construction.

Other effects which affect insulation are vaporization of pyrolysis products, e.g., from FR finishes for cotton, and recondensation on the heat sensor or skin,[191] and fabric–skin contact which is less when the fabrics have protruding fibers or a rough weave than when the fabric surface is smooth.[192] Since fabric thickness is the primary factor for insulation, the fibers for protective fabrics can be chosen on the basis of their resistance to ignition and heat shrinkage, as well as price, comfort, appearance, etc.; e.g., relatively cheap modacrylics have a LOI of about 30 but shrink and char at lower temperatures than FR cotton and wool.[193] At a given heat load, e.g., 0.30 J/cm^2 sec for 1 min, the char formed by wool fabrics is much weaker than that formed by FR cotton, and Nomex is merely discolored; polyester and nylon form melt holes.

A proposed ASTM method for thermal protective clothing[194] exposes textiles to a gas flame delivering 8.4 W/cm^2. The specimens are horizontal and can be held rigidly, or, if heat shrinkage is of interest, loosely. A copper sensor is 3.2 mm above the upper surface of the specimen. The time to incipient second degree, based on the "Stoll criteria",[195] is determined.

A British test[196] employs a heat source consisting of 150-ml hexane floating on water on one side of the fabric specimen and a copper sensor on the other. The sensor is pressed against the fabric specimen by spring loading. The time until the sensor registers a 25°C temperature rise is noted.

A study of the radiative characteristics of firefighters' turn-outs coats[197] showed that color has little effect on their protective value, but aluminized outer layers increased protection significantly. However, such reflective surfaces tended to loose their effectiveness when they became dirty.

Heat transfer through fabrics from flame contact and from a purely radiative source were compared.[198] The ranking of the fabrics was roughly the same. Similar rankings were also obtained when fabrics were exposed to mixed radiative–convective heat and to purely radiation heat.[192]

For many situations, protective clothing must also facilitate rolling off of molten metal.[199] Steel rolls off cotton with FR finishes, but aluminum requires special formulation.

11. *Thermal Behavior of Textile Materials*

This section summarizes the behavior of textile fiber materials when exposed to small or moderate size heat sources. The behavior of high temperature fibers is beyond the scope of this chapter; they do not ignite from the ignition sources considered here, and shrink only slightly, if at all.[128]

Table 2 attempts to classify the various fibers according to melt and shrinkage, char, and burning behavior. Wool is in the class of fibers which

TABLE 2
Melting and Burning Behavior of Textile Materials

Char and burn: cellulosic fibers, e.g., cotton, linen, jute, viscose rayon, etc.; acrylics; wool (burns to a lesser degree).
Shrink ahead of flame, melt, ablate, but burn when held in place: thermoplastic fibers, e.g., acetate, nylon, polyester, olefins, triacetate.
Do not ignite from small ignition sources but shrink to various degrees and char: PVC, matrix, modacrylic, aramid.

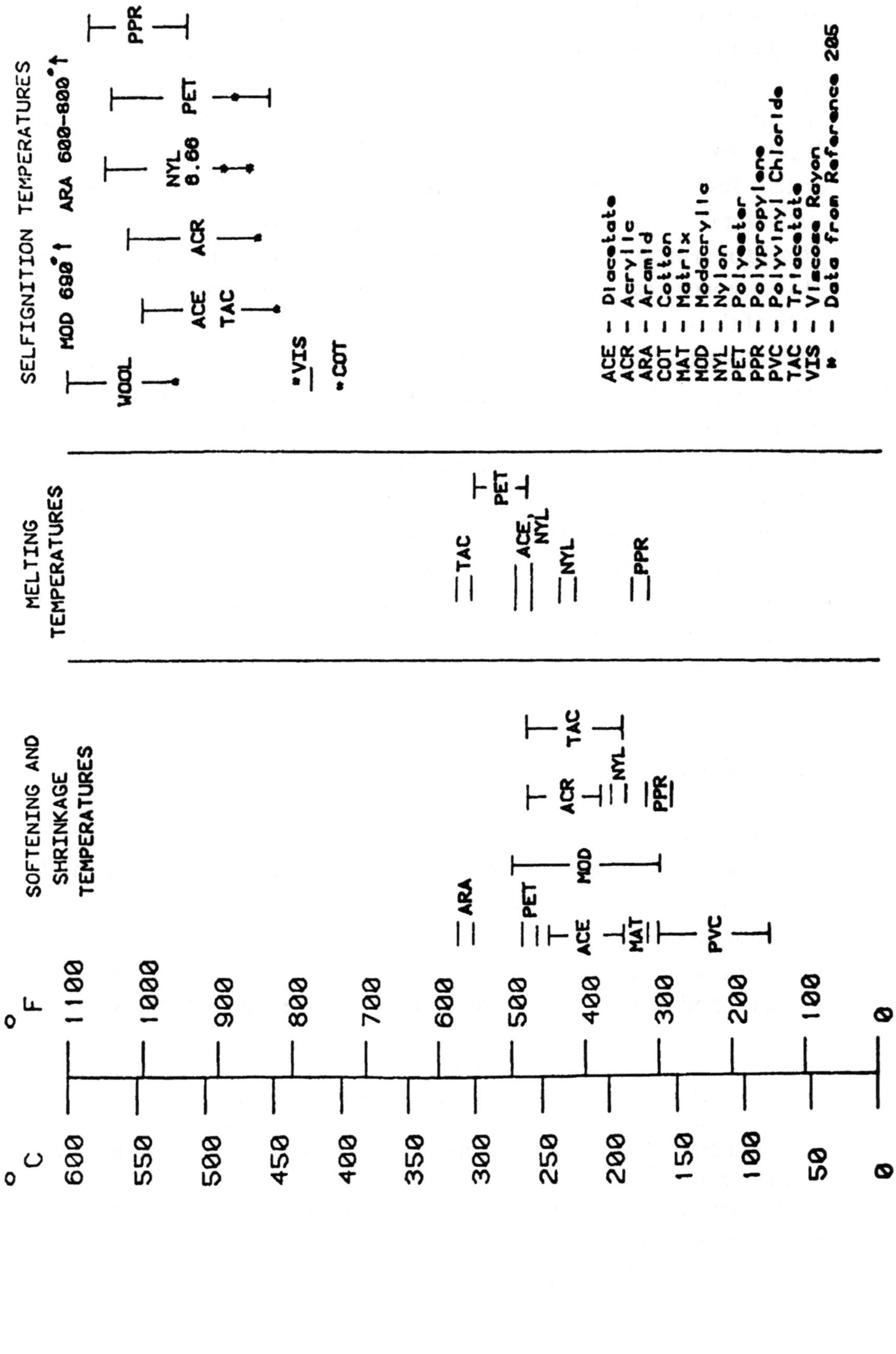

FIGURE 1. Some thermal properties of textile materials.

char and burn, but it does so to a lesser degree and at a slower rate than the other fibers in this class, acrylics and cellulosics. Similarly, most thermoplastics in fabric form melt away from the flame unless held in place by, e.g., a seam or a finish. However, in the case of acetate and triacetate, woven fabrics often ignite and burn vigorously, while knitted fabrics tend to shrink from the flames and do not ignite and burn as readily.

Figure 1 shows the ranges listed in the literature for the softening and shrinkage temperatures of various fibers, as well as their melting and self-ignition temperatures.[200–207] The methods for obtaining the various values are often not specified, and may differ sufficiently to account for some of the large ranges shown. In addition, some of these ranges are undoubtedly due to the large differences in polymer structure of the commercial fibers in any one generic class. One laboratory determined ignition temperatures for most commercial fibers.[205] Its results are indicated by a special symbol; in general, these results were on the lower end of the range of values found in the literature, with the exception of cotton. The procedure used consisted of forming 250-mg balls of the fibers and placing them into a furnace at a given temperature. Ignition or no ignition were noted. The same investigators determined the temperature at which 50% of the fiber ball had decomposed. They obtained the same values for fiber balls consisting of 50/50 blends of the various fibers with cotton.

12. *References*

1. C7 Publications Committee, Textile Flammability, American Association of Textile Chemists and Colorists, Research Triangle Park, North Carolina (1975).
2. Title 16, Part 1615-Standard for the Flammability of Children's Sleepwear: Sizes 0 through 6X (FF 3-71); Part 1616 Sizes 7 through 14 (FF 5-74), Code of Federal Regulations, Office of the Federal Register, General Services Administration, Washington, D.C., pp. 365–417 (1978).
3. Title 16, Part 1632—Standard for the Flammability of Mattresses (and Mattress Pads) (FF 4-72), Code of Federal Regulations, Office of the Federal Register, General Services Administration, Washington, D.C., pp. 434–469 (1978).
4. Method of Test for Combustion Resistance of Mattresses, Cigarette Test, 27.7-M77, Canadian Government Specification Board, Ottawa, Ontario (1977).
5. Test Development Subcommittee, Apparel Products Flammability Committee, American Association of Apparel Manufacturers, Development of a semirestrained vertical fabric flammability test concept, *Text. Chem. Color* **5**, 215–221 (1973).
6. G. M. Richardson, ASTM flammability test programs for textiles, in Proceedings of the 1975 Symposium on Textile Flammability, pp. 49–63, LeBlanc Research Corporation, East Greenwich, Rhode Island 02818 (1975).
7. T. A. Pressley, K. W. Fincher, and R. J. Rowlands, Textile flammability testing, *Text. J. Austr.* **48**(2), 1, 18–21, 38–40 (1973).
8. L. J. Sharman and H. Tovey, Current status and national priorities for flammable fabric standards, in Proceedings of the Sixth Annual Meeting, Information Council for Fabric Flammability, 264–306, New York (1972).

9. Title 16, Part 1615, Final rules, *Fed. Regist.* **43**, 4849–4855 (February 6, 1978).
10. E. Braun, V. B. Cobble, S. Helzer, J. F. Krasny, R. D. Peacock, and A. K. Stratton, Back-up Report for the Proposed Standard for the Flammability of General Wearing Apparel, NBSIR 76-1072, National Bureau of Standards, Washington, D.C., (1976).
11. K. A. Jakes, E. F. Schneider, B. F. Smith, and S. M. Spivak, Factors affecting seam flammability, *Text. Chem. Color.* **6**, 53–56 (1974).
12. D. Hyman, B. F. Smith, and R. Dardis, Cost-benefit analysis of flammability for children's sleepwear and clothing: Sizes 7–14, in Proceedings of the Tenth Annual Meeting, Information Council on Fabric Flammability, pp. 297–304, New York (1976).
13. S. Bolieu, Children's sleepwear flammability standards: Have they worked? in Proceedings of the Eleventh Annual Meeting, Information Council on Fabric Flammability, pp. 1–5, New York (1977).
14. G. F. Crikelair, A win for the team, in Proceedings of the Ninth Annual Meeting, Information Council for Fabric Flammability, pp. 218–233, New York (1975).
15. Flame Resistance of Cloth; Vertical Method 5903.2, Federal Test Method Standard No. 191, General Services Administration, Washington, D.C. (1971).
16. CPAI-84, A Specification for Flame Resistant Materials Used in Camping Tentage, Canvas Products Association International, St. Paul, Minnesota (1976).
17. Title 14, Part 25—Airworthiness standards: Transport category airplanes, Section 25.853, *Fed. Regist.* **37**, 3792–3794 (February 24, 1972).
18. Flame Resistance of Cloth; Vertical, Field Method 5904, Federal Test Method Standards No. 191, General Services Administration, Washington, D.C. (1968).
19. Flame Resistance of Material: High Heat Flux Flame Contact, Method 5905, Federal Test Method Standard No. 191, General Services Administration, Washington, D.C. (1971).
20. Standard Methods of Fire Tests for Flame-Resistant Textiles and Films, NFPA 701–1976, National Fire Protection Association, Boston, Massachusetts (1976).
21. Tentative Test Method for Flammability of Apparel by Semi-Restraint Method, ASTM D 3659-77 T, American Society for Testing and Materials, Philadelphia (1978).
22. Determination of the Burning Behavior-Vertical Method; Ignition by Application of Flame to Base of Specimen, DIN 53906, Deutsches Institut für Normung (1974).
23. Classification of the Burning Behavior of Textile Products: Curtain and Drapery Materials, DIN 66082, Deutsches Institut für Normung (1975).
24. Method of Test for Flame Resistance 27.1-M 77, Vertical Burning Test, Canadian Government Specification Board, Ottawa, Ontario (1977).
25. Method of Test for Flame Resistance 27.3–M77, Rate-of-Burning Test, Canadian Government Specification Board, Ottawa, Ontario (1977).
26. Flame Resistance of Cloth: Horizontal, Method 5900, Federal Test Method Standard No. 191, General Services Administration, Washington, D.C. (1968).
27. Australian Standard 1176, Parts 1 to 3—1976, Method of Test for Combustion Characteristics of Textile Materials, Standards Association of Australia, North Sidney, New South Wales (1976).
28. Australian Standard 1248–1976, Fabrics for Domestic Apparel of the Low Fire Hazard Type, Standards Association of Australia, North Sidney, New South Wales (1976).
29. Australian Standard 1249, Parts 1 and 2—1976, Part 3—1977, Children's Night Clothes Having Reduced Fire Hazard, Standards Association of Australia, North Sidney, New South Wales (1976, 1977).
30. Australian Standard 1182—1972, Size Coding Scheme for Infant's and Children's Clothing (Underwear and Outerwear), Standards Association of Australia, North Sidney, New South Wales (1972).
31. Australian Standard 1989—1976, Classification and Labelling of Children's Night Clothes for Fire Hazard, Standards Association of Australia, North Sidney, New South Wales (1976).

32. P. G. Gordon, Standards for the fire hazard of clothing: The Australian experience, *Fire Mater.* **2**, 163–172 (1978).
33. Methods of Test for Flammability of Vertically Oriented Textile Fabrics and Fabric Assemblies Subjected to a Small Igniting Flame, BS 5438: 1976, British Standards Institution, London (1976).
34. J. G. Feikema, The combustion rates of various textiles, *Melliand Textilber.* **54**, 179–184, (1973).
35. E. B. Nielsen and H. R. Richards, A proposed method for measuring rate-of-burn, *Text. Chem. Color.* **1**, 270–277 (1969).
36. D. I. Lawson, Fire accidents—The contribution of some textiles, *Research* **11**, 126–133 (1958).
37. A. Vickers, J. Krasny, and H. Tovey, Some apparel fire hazard parameters, in Proceedings of the Seventh Annual Meeting, Information Council for Fabric Flammability, pp. 205–226, New York (1973).
38. L. D. Moore, Full Scale Burning Behavior of Curtains and Draperies, NBSIR 78-1448, National Bureau of Standards, Washington, D.C. (1978).
39. Title 16, Part 1610—Standard for the Flammability of Clothing Textiles (formerly CS 191-53), Code of Federal Regulations, Office of the Federal Register, General Services Administration, Washington, D.C. (1978).
40. Standard for the Classification of the Flammability of Wearing Apparel, NFPA 702-1975, National Fire Protection Association, Boston, Massachusetts (1976).
41. Burning Rate of Cloth; 45° Angle, Method 5908, Federal Test Method Standard No. 191, General Services Administration, Washington, D.C. (1968).
42. Method of Test for Flame Resistance 27.5, Canadian Government Specification Board, Ottawa, Ontario (1977).
43. Method of Test for Flame Resistance 27.2-M77, Surface Burning Test, Canadian Government Specification Board, Ottawa, Ontario (1977).
44. Determination of the Burning Behavior—45° Method, DIN 54335, Deutsches Institut für Normung (1973).
45. H. Mori, Teijin's flame retardant fiber tetoron extar, in Proceedings of the 1975 Symposium on Textile Flammability, LeBlanc Research Corporation, East Greenwich, Rhode Island, pp. 125–154 (1975).
46. Tentative Test Methods for Flammability of Textile Materials, D3411-75T, American Society for Testing and Materials, Philadelphia (1975).
47. C. H. Irvine, Modified children's sleepwear test chamber for evaluation of relative flammability of garment fabrics, in Proceedings of the Eleventh Annual Meeting, Information Council on Fabric Flammability, pp. 94-106, New York (1977).
48. Burning Rate of Cloth; 30° Angle, Method 5910, Federal Test Method Standard No. 191, General Services Administration, Washington, D.C. (1968).
49. Burning Rate of Cloth; Horizontal, Method 5906, Federal Test Method Standard No. 191, General Services Administration, Washington, D.C. (1968).
50. Motor Vehicle Safety Standard No. 302, Flammability of Interior Materials—Passenger Cars, Multipurpose Passenger Vehicles, Trucks, and Buses, Part 571, Code of Federal Regulations, Office of the Federal Register, General Services Administration, Washington, D.C. (November 28, 1975).
51. Determination of the Flame Propagation Speed of Burnable Materials: Horizontal Method; Ignition by Applying the Flame to the Edge of the Material, DIN 54333, Deutsches Institut für Normung (1973).
52. G. Stamm, Heutiger Stand der Brenn- und Entflammbarkeitsprüfung von Textilien, *Textilveredlung* **6**, 103–112 (1971).
53. Determination of Burning Behavior—Test Method by Semicircle, DIN 54331, Deutsches Institut fur Normung (1974).

54. E. Meherki and M. Romanowski, The measurement of fire in vertically suspended fabric: Upward Acceleration of Flame Front, *Text. Res. J.* **35**, 500–504 (1975).
55. J. W. Weaver, Rate of burning of apparel fabrics, *Text. Chem. Color.* **8**, 176–181 (1976).
56. R. Aenishanslin and F. Mayer, Physikalische Einflussfaktoren auf das Brennverhalten von Geweben, *Textilveredlung* **11**, 329–335 (1976).
57. E. B. Neilsen and H. R. Richards, Measuring the rate-of-burn of thermoplastic fabrics, *Text. Chem. Color.* **2**, 35–40 (1970).
58. F. H. Holmes, Flammability Testing of Fabrics, International Symposium, Fire Safety of Combustible Materials, Edinburgh, pp. 317–323 (1975).
59. C. D. Ward and S. M. Jaeckel, Measurement of the flammability of apparel fabrics, *J. Text. Inst.* **67**, 309–318 (1976).
60. B. Miller, J. R. Martin, B. C. Goswami, and C. H. Meiser, Jr., The effects of moisture on the flammability characteristics of textile materials, *Text. Res. J.* **45**, 329–337 (1975).
61. B. Miller, J. R. Martin, and C. H. Meiser, Jr., Measurement of burning rates of unrestrained fabrics, *Text. Chem. Color.* **7**, 68–72 (1975).
62. P. Durbetaki and M. L. Teague, Fabric ignition studies with a microburner, in Proceedings of the Eleventh Annual Meeting, Information Council on Fabric Flammability, New York, pp. 141–159 (1977).
63. C. C. Ndubizu and P. Durbetaki, Modeling radiative ignition in air, *Fire Res.* **1**, 281–290 (1977/78).
64. S. Backer, G. C. Tesoro, T. Y. Toong, and N. A. Moussa, *Textile Fabric Flammability, Part III: Ignition*, MIT Press, Cambridge, Massachusetts (1975).
65. Method of Test for Flame Resistance 27.4-M77, Ease of Ignition Test, Canadian Government Specification Board, Ottawa, Ontario (1977).
66. Determination of the Burning Behavior—Ignition Time, Edge Ignition, DIN 54334, Deutsches Institut für Normung (1974).
67. Proposed Standard for the Flammability of General Wearing Apparel, National Bureau of Standards, Washington, D.C. (1976).
68. A. L. Fisher, L. Fourt, and J. F. Krasny, The effect of fabric and garment parameters on flammability, Proceedings of the Fifth Annual Meeting, Information Council on Fabric Flammability, New York, pp. 119–137 (1971).
69. L. B. Miles, An investigation of the flammability hazard of apparel fabrics, in Proceedings of the 1977 Symposium on Textile Flammability, LeBlanc Research Corporation, East Greenwich, Rhode Island, pp. 38–74 (1978).
70. C. T. Webster, H. G. H. Wraight, and P. H. Thomas, Heat-transfer from burning fabrics, *J. Text. Inst. Trans.* **53**, T29–37 (1962).
71. J. A. Pete, Evaluation of fyrol treated fabrics using the mushroom apparel flammability tester, *Text. Chem. Color.* **8**(8), 22–24 (August 1977).
72. E. A. Zawistowski, J. F. Krasny, E. Braun, R. Peacock, and N. Williams, The Measurement of Fabric Flammability Parameters in Experiments Simulating Human Movement in Burn Accidents, NBSIR 77-1236, National Bureau of Standards, Washington, D.C. (1977).
73. J. E. Smith, A review of the NBS mushroom apparel flammability tester (MAFT), in Proceedings of the 1976 Symposium on Textile Flammability, pp. 63–88, LeBlanc Research Corporation, East Greenwich, Rhode Island (1976).
74. L. Segal and G. L. Drake, Jr., Performance of a mushroom apparel flammability tester in evaluating "partially flame retardant" cotton flannelette, *J. Flame Retardant Chem.* **6**, 3–13 (1979).
75. L. Segal and G. L. Drake, Jr., Mushroom flammability tester, *Text. Chem. Color.* **9**, 244–245 (1977).
76. E. A. McCullough and C. J. Noel, Flammability characteristics of layered fabric assemblies, in Proceedings of the Twelfth Annual Meeting, Information Council on Fabric Flammability, New York, pp. 175–184 (1978).

77. B. Miller, J. R. Martin, C. H. Meiser, Jr., and M. Gargiullo, The flammability of polyester–cotton mixtures, *Text. Res. J.* **46**, 530–538 (1976).
78. B. Miller, A new concept for monitoring the burning behavior of fabrics, in Proceedings of the Tenth Annual Meeting, Information Council on Fabric Flammability, New York, pp. 198–208 (1976).
79. J. R. Martin and B. Miller, The thermal and flammability behavior of polyester–wool blends, *Text. Res. J.* **48**, 97–103 (1978).
80. B. Miller and C. H. Meiser, Jr., Heat emission from burning fabrics; potential harm ranking, *Text. Res. J.* **48**, 238–243 (1978).
81. M. P. Chouinard, D. C. Knodel and H. W. Arnold, Heat transfer from flammable fabrics, *Text. Res. J.* **43**, 166–175 (1973).
82. J. F. Krasny and A. L. Fisher, Laboratory modeling of garment fires, *Text. Res. J.* **43**, 272–283 (1973).
83. G. B. Arnold, A. Fisher, and G. Frohnsdorff, Hazards from Burning Garments, Gillette Research Institute, Rockville, Maryland 20850 (1974).
84. J. R. Bercaw, Cooperative program on general apparel flammability: status report, Proceedings of the Eleventh Annual Meeting, Information Council on Fabric Flammability, New York, pp. 175–190 (1977).
85. J. R. Bercaw and K. G. Jordan, Injury from burning garments, in Proceedings of the Eighth Annual Meeting, Information Council on Fabric Flammability, New York, pp. 204–213 (1974).
86. A. W. Meierhoefer, E. Braun, J. F. Krasny, and R. D. Peacock, Fire Accident Simulations with Apparel Fabrics, NBSIR 79-1755, National Bureau of Standards, Washington, D.C. (1979).
87. F. Galil, Flammability of modacrylic/acrylic fibers. Part I: Effect of the blend ratio, *Text. Chem. Color.* **9**, 240–243 (1977); Part II: Effect of fabric construction, *Text. Chem. Color.* **10**, 35–38 (1978).
88. F. Galil, Predicting garment flammability, *Text. Chem. Color.* **11**, 66–72 (1979).
89. K. H. Umbach, Brennverhalten von Textilien aus PES/Baumwolle und 100% Baumwolle, *Zentral. Arbeitsmed. Arbeitsschutz Prophylaxe* **27**, 32–37 (1977).
90. G. F. Crikelair, F. Agate, and A. Bowe, Gasoline flammable and nonflammable clothing studies, *Pediatrics* **50**, 585–594, (1976).
91. B. B. Hibbard and J. F. Krasny, Preliminary flammability experiments with fabrics contaminated with gasoline, in Proceedings of the Twelfth Annual Meeting, Information Council on Fabric Flammability, New York (1978).
92. L. B. Buchbinder, Human behavioral patterns vs injury severity for apparel fire victims, in Proceedings of the Seventh Annual Meeting, Information Council for Fabric Flammability, New York, pp. 236–253 (1973).
93. G. R. Bhat, I. Block, K. Yeh, and S. M. Spivak, Extinguishability of flammable textiles. Part I. Statement of the problem, *J. Consumer Product Flammability* **3**, 155–165 (1976).
94. L. Miles, G. R. Bhat, R. L. Sun, K. Yeh, and S. M. Spivak, Extinguishability of flammable textiles. Part II. Effects of ambient oxygen concentration, *J. Consumer Product Flammability* **3**, 166–192 (1976).
95. L. B. Miles, G. R. Bhat, R. I. Sun, K. Yeh, and S. M. Spivak, Extinguishability of flammable textiles. Part III. Effects of cooler, inert bodies, *J. Consumer Product Flammability* **5**, 135–162 (1978).
96. P. P. Pothoff, R. L. Sun, F. W. Derrick, and S. M. Spivak, Extinguishability of flammable textiles. Part IV. Effects of air flow, *J. Consumer Product Flammability* **6**, 261–287 (1979).
97. J. F. Krasny and J. B. Francis, Ranking extinguishability of burning fabrics, *Text. Chem. Color.* **6**, 135–136 (1974).
98. B. Miller, B. C. Goswami, and R. Turner, The concept and measurement of extinguishability as a flammability criterion, *Text. Res. J.* **43**, 61–67 (1973).

99. R. B. Le Blanc and J. P. DiCarlo, Tests for ease of extinguishment, in Proceedings of the 1977 Symposium on Textile Flammability, pp. 39–65, LeBlanc Research Corporation, East Greenwich, Rhode Island (1977).
100. A. McQuade, Meeting of Research Committee, Information Council on Fabric Flammability, Natick, Massachusetts (May 31, 1973).
101. P. P. Carfagno and J. F. Pacheco, Laundering variables and their effect on flame retardant fabrics, University of Utah Polymer Conference Series, Salt Lake City (1973).
102. R. S. Merkel, W. L. Mauldin, M. A. Zentner, V. Ahrens, C. Avery, B. Smith, and S. Spivak, Effects of wear and laboratory launderings on flammability of four women's nightgown fabrics, in Proceedings of the Twelfth Annual Meeting, Information Council on Fabric Flammability, New York, pp. 62–78 (1978).
103. A. Adler and P. Stensby, Factors influencing flammability of flame retardant fabrics, in *Text. Chem. Color.* **7**, 7–15 (1975).
104. R. J. Brysson, W. A. Reeves, A. M. Walker, and B. Piccolo, Durability of flame retardants influenced by detergent selection and laundering variables, in Proceedings of the Fifth Annual Meeting, Information Council on Fabric Flammability, pp. 138–164, New York (1971).
105. J. F. Pacheco and P. P. Carfagno, How laundering practices influence the flame retardancy of fabrics, *Text. Chem. Color.* **4**, 255–259 (1972).
106. T. C. Defosse and P. P. Carfagno, The effect of laundering practices on the flame retardance of fabrics, *Text. Chem. Color.* **6**, 25–29 (1974).
107. P. K. Woodward, Effective Labeling of Flame Resistant Apparel, The Clorox Company, Oakland, California.
108. T. J. Swihart and P. E. Campbell, How silicones affect fabric flammability, *Text. Chem. Color.* **6**, 109–112 (1974).
109. H. N. Simpson and C. C. Silvernale, How laundering with fabric softeners affects flame retardant cotton fabrics, *Am. Dyes. Rep.* **58**, 36–42 (1976).
110. A. L. Bullock and M. E. Carter, Effect of soiling with milk and urine on fire retardant cotton fabrics, *Am. Dyes. Rep.* **57**, 21–24 (1975).
111. O. P. Beckwith, The pajama game—Post FR, in Proceedings of the 1975 Symposium on Textile Flammability, pp. 81–92, LeBlanc Research Corporation, East Greenwich, Rhode Island (1975).
112. H. N. Simpson and M. A. Phelps, How glycerine hand lotions and baby oil affect fire retardant fabrics, *Am. Dyes. Rep.* **57/8**, 69–70 (1975).
113. I. Hardin, A. J. Treece, S. H. Zeronian, R. L. Galbraith, and B. R. Farthing, The effect of FR finishes on the soiling and soil removal characteristics of cotton and cotton/polyester blend fabrics, *J. Consumer Product Flammability* **4**, 121–155 (1977).
114. B. M. Reagan and C. M. Chiasson, Photodegration of flame-water-weather-and mildew resistant cotton tentage fabric, *Text. Res. J.* **47**, 637–646 (1977).
115. A. J. Treece, J. L. Bare, and J. O. Mundt, Mildew stains on FR-treated cotton fabrics, *Text. Chem. Color.* **9**, 226–230 (1977).
116. B. L. Slaten, Effects of weathering and atmospheric pollutants on cotton fabric and cotton fabric treated with selected flame retardants, No. 75-9498, Xerox University Microfilms, Ann Arbor, Michigan (1975).
117. L. Segal, C. E. Morris, and G. L. Drake, Jr., Effects of hospital laundering and sterilization on FR fabrics, I. A preliminary study on cotton flannelette, Book of Papers, National Conference of the American Society of Textile Chemists and Colorists, pp. 205–212 (1975).
118. L. Segal, C. E. Morris, and G. L. Drake, Jr., Effect of hospital laundering and sterilization on FR fabrics, II. Cotton and cotton–polyester fabrics laundered with bleach, *Am. Dyes. Rep.* **66/1**, 39–44, 52–54 (January 1977).
119. C. E. Morris, L. Segal, and G. L. Drake, Jr., Effect of hospital laundering and sterilization on FR fabrics, III. man-made fibers laundered with bleach, *Am. Dyes. Rep.* **66/10**, 54–59, 79 (October 1977).

120. C. E. Morris, L. Segal, and G. L. Drake, Jr., Effects of hospital laundering and sterilization on FR fabrics, IV. Fabrics laundered without bleach, *Am. Dyes. Rep.* **67/4**, 30–39, 55 (April 1978).
121. M. Vaishnav, Drycleaning Procedure for Inclusion in Fabric Flammability Standards, Center for Fire Research, National Bureau of Standards, Washington, D.C. (1975).
122. J. R. Martin and B. Miller, Effects of laundering on flame retardance, in Proceedings of the Tenth Annual Meeting, Information Council on Fabric Flammability, pp. 221–227, New York (1976).
123. C. P. Fenimore and F. J. Martin, Candle-type test for flammability of polymers, *Mod. Plast.* **44**, 41–48, 92 (1966).
124. Palmetto Section, Flammability testing: Measurements and relationships, *Text. Chem. Color.* **4**, 71–76 (1972).
125. J. F. Hendrix, G. L. Drake, Jr., and W. A. Reeves, Some factors affecting fabric flammability as measured by the oxygen index (OI), *Text. Chem. Color.* **5**, 144–148 (1973).
126. B. Miller and C. H. Meiser, Jr., Steady state burning of textiles in flowing O_2/N_2 mixtures, *Text. Chem. Color.* **2**, 205–208 (1970).
127. G. C. Tesoro, Flame resistant polyester/cotton fabrics, *J. Appl. Polym. Sci.* **24**, 1073–1084 (1977).
128. N. J. Abbott and S. Schulman, Protection from fire: Nonflammable fabrics and coatings, *J. Coated Fabr.* **6**, 48–64 (July 1976).
129. L. Segal, C. E. Morris, and M. M. Smith, A statistical study of the oxygen index measurement, *Text. Res. J.* **46**, 620–621 (1976).
130. K. A. Jakes, B. F. Smith, and S. M. Spivak, A modified bottom ignition oxygen index test for determination of the flammability of sewing threads, *Text. Res. J.* **44**, 398–399 (1974).
131. R. Sanders, A small scale burning test for shag carpets, *Text. Chem. Color.* **5**, 48–51 (1973).
132. M. Day and D. M. Wiles, Carpet flammability: An oxygen index modification to the pill test, *J. Fire Flammability* **4**, 165–173 (1973).
133. Relationships of flammability measurements, *Text. Chem. Color.* **1**, 540–548 (1968).
134. D. L. Hopkins and D. R. Buchanan, The effect of reactive dyes on the flammability and physical properties of polyester/cotton fabrics, *Text. Res. J.* **48**, 348–354 (1978).
135. J. D. Timpa, L. Segal, and G. L. Drake, Jr., Dyes and flame-retardant cotton textiles, *J. Fire Flammability Flame Retard. Chem.* **2**, 83–93 (1975).
136. L. Segal, J. D. Timpa, and G. L. Drake, Jr., The effect of dyes on the flammability of cotton fabric, *Text. Res. J.* **44**, 839–846 (1974).
137. Delaware Valley Section, Effect of disperse dye on certain flammability characteristics of polyester, *Text. Chem. Color.* **10**, 252–259 (1978).
138. G. C. Tesoro and C. H. Meiser, Jr., Some effects of chemical composition on the flammability behavior of textiles, *Text. Res. J.* **40**, 430–436 (1970).
139. G. C. Tesoro and J. Rivlin, Flammability behavior of experimental blends, *Text. Chem. Color.* 3, 156–160 (1971).
140. L. Segal, C. E. Morris, and G. L. Drake, Jr., The effect of halogen-containing acid dyes on the flammability of cotton flannelette, *J. Fire Retard. Chem.* **5**, 183–195 (1978).
141. Notice of finding that flammability standards or other regulations may be needed and instituton of proceedings, (15 CFR Part 7), *Federal Register*, 35/112, 46 (June 10, 1970).
142. Title 16, Part 1630—Standard for the Surface Flammability of Carpets and Rugs (FF 1-70), Code of Federal Regulations, Office of the Federal Register, General Services Administration, Washington, D.C., pp. 417–424 (1978).
143. Title 16, Part 1630—Standard for the Surface Flammability of Small Carpets and Rugs (FF 2-70), Code of Federal Regulations, Office of the Federal Register, General Services Administration, Washington, D.C., pp. 424–434 (1978).
144. Method 27.6-M77, Method of Test for Flammability—Methanamine Tablet Test—for Textile Floor Coverings, Canadian Government Specification Board, Ottawa, Ontario (1977).

145. K. Tu and S. Davis, Flame Spread of Carpet Systems Involved in Room Fires, NBSIR 76-1013, National Bureau of Standards, Washington, D.C., (1976).
146. Determination of Flammability of Floor Coverings (Hot Metal Nut Method), British Standard 4790:1972, British Standards Institution, London (1972).
147. Assessment and Labeling of Textile Floor Coverings Tested to BS 4790, BS 5287:1976, British Standards Institution, London (1976).
148. Testing of Textiles: Determination of the Burning Behavior of Textile Floor Coverings, DIN 54332, Deutsches Institut für Normung (1975).
149. Classification of Burning Behavior of Textile Products; Textile Floor Coverings, DIN 66081, Deutsches Institut für Normung (1976).
150. E. C. Thomm, Effects of carpet variables on the methanamine pill test, *J. Fire Flammability* **4**, 197–209 (1973).
151. M. Day, M. T. Mitton, and D. M. Wiles, An Experimental procedure for determining the flammability of carpets using the TRI flammability analyzer, *Text. Res. J.* **34**, 5–11 (1974).
152. Standard Method of Test for Critical Radiant Flux of Floor Covering Systems Using a Radiant Heat Energy Source, Part 1, NFPA 253-1978, National Fire Protection Association, Boston, Massachusetts (1978), also ANSI/ASTM E 648–78, American Society for Testing and Materials, Philadelphia (1978).
153. I. A. Benjamin and S. Davis, Flammability Testing for Carpet, NBSIR 78-1436, National Bureau of Standards, Washington, D.C. (1978).
154. S. Alderson and L. Breeden, Evaluation of the Fire Performance of Carpet Underlayments, NBSIR 76-1018, National Bureau of Standards, Washington, D.C. (1976).
155. J. Quintiere, The application and interpretation of a test method to determine the hazard of floor covering fire spread in building corridors, in International Symposium, Fire Safety of Combustible Materials, Edinburgh, pp. 355–366a (1975).
156. Standard Methods of Test for Surface Burning Characteristics of Building Materials, ASTM E 84-70, American Society for Testing and Materials, Philadelphia (1970); also NFPA 255-1972, National Fire Protection Association, Boston, Massachusetts (1972).
157. T. G. Lee and C. Huggett, Interlaboratory evaluation of ASTM E 84-70 tunnel test applied to floor coverings, *J. Test. Eval.* **3**, 3–13 (1975).
158. Standard Method of Test for Surface Burning Characteristics of Flooring and Floor Covering Materials, B.54.9-1972, Canadian Government Specification Board, Ottawa, Ontario (1972).
159. H. S. Engerman, Floor covering system—A test for flame propagation index, *J. Fire Flammability* **2**, 11–35 (1971); also Standard on the Chamber Furnace Test Method for Measuring the Surface Flame Propagation Characteristics of Flooring and Floor Covering Materials, Part II, NFPA 254-1978, National Fire Protection Association, Boston, Massachusetts (1978).
160. R. E. Frye, A. K. Vickers, and E. A. Tyrrell, Analysis of Accident Data on Mattress Fires, NBS Report 10820, National Bureau of Standards, Washington, D.C. (1972).
161. C. J. Hilado and P. A. Huttlinger, A bibliography of published information on smoldering combustion, *J. Consumer Product Flammability* **5**, 260–264 (1978).
162. V. Babrauskas, Combustion of Mattresses Exposed to Flaming Ignition Sources, Part 1. Full-Scale Tests and Hazard Analysis, NBSIR 77-1290, National Bureau of Standards, Washington, D.C. (1977).
163. C. J. Hilado, ed., *Bedding and Furniture Materials*, Technomic Publishing, Westport, Connecticut (1976).
164. N. B. Knoepfler, J. P. Newmyer, and P. A. Koenig, Backcoated mattress tickings as a factor in meeting mattress flammability standard FF 4-72, Part 1, *J. Fire Flammability, Consumer Product Flammability* **2**, 70–83; Part 2, Supplement 2, 123–139 (1975).
165. N. B. Knoepfler, J. P. Madasci, and W. T. Gentry, Boric acid treated mattresses resist both open flame and cigarette ignition, *J. Consumer Product Flammability* **5**, 26–40 (1978).

166. C. H. Yuill, The life hazard of bedding and upholstery fires, *J. Fire Flammability* **1**, 312–323 (1970).
167. C. Anolick, L. R. Perkins, C. W. Stewart, J. R. Galloway, and W. Hellyer, Flammability characteristics of healthcare mattresses, *J. Consumer Product Flammability* **5**, 168–198 (1978).
168. P. S. Schaenman and H. Tovey, Upholstered Furniture Fires; The Magnitude of the Problem, National Fire Data Center, U.S. Fire Administration, Washington, D.C. (December 20, 1978).
169. Part 1633—Proposed Standard for the Flammability (Cigarette Ignition Resistance) of Upholstered Furniture, U.S. Consumer Products Safety Commission, Washington, D.C. (1978).
170. D. J. Donaldson, H. H. St. Mard, and R. J. Harper, Jr., Imparting smolder-resistance to cotton upholstery fabrics, *Text. Res. J.* **49**, 185–190 (1979).
171. Anonymous, Sulphur vs cigarette ignition, *Am. Bedding Mag.* **105/6**, 23–25 (June 1977).
172. J. J. Loftus, Back-up Report for the Proposed Standard for the Flammability (Cigarette Ignition Resistance) of Upholstered Furniture, PFF 6-76, NBSIR 78-1438, National Bureau of Standards, Washington, D C. (1978).
173. R. Dardis and R. Thompson, Cost-benefit analysis of the proposed flammability standard for upholstered furniture, in Proceedings of the 1978 Symposium on Textile Flammability, pp. 93–102, LeBlanc Research Corporation, East Greenwich, Rhode Island (1978).
174. S. G. Helzer, B. Buchbinder, and F. Offensend, Decision Analysis of Strategies for Reducing Upholstered Furniture Fire Losses, TN1101, National Bureau of Standards, Washington, D.C. (1979).
175. How to Comply with the UFAC Voluntary Action Program, Upholstered Furniture Action Council, High Point, North Carolina (1979).
176. Tests for the Ignitability of Upholstered Seating, DD 58:1978, British Standards Institution, London (1978).
177. California Bureau of Home Furnishings Technical Bulletin 116, Requirements, Test Procedure and Apparatus for Testing the Flame Retardance of Upholstered Furniture, Sacramento, California (1975)
178. California Bureau of Home Furnishings Technical Bulletin 117, Requirements, Test Procedures and Apparatus for Testing the Flame Retardancy of Filling Materials Used in Upholstered Furniture, Sacramento, California (1975).
179. G. H. Damant, A bibliography of published information on furniture flammability, *J. Consumer Product Flammability* **5**, 245–249 (1978).
180. G. H. Damant and M. A. Young, Smoldering characteristics of fabrics used as upholstered furniture coverings, *J. Consumer Product Flammability* **4**, 60–113 (1977).
181. G. H. Damant and M. A. Young, Flammability classification of fabrics used as upholstered furniture coverings, *J. Consumer Product Flammability* **4**, 329–345 (1977).
182. C. J. Hilado, H. J. Cumming, and A. Furst, Fire Safety in the Home: Flash Fire Propensity of Some Upholstery Fabrics, Laboratory Report No. SP-77-5, Bureau of Home Furnishings, Department of Consumer Affairs, Sacramento, California (1977).
183. C. J. Hilado and D. P. Brauer, A compilation of relative toxicity data on furniture upholstery fabrics, *J. Consumer Product Flammability* **5**, 250–259 (1979).
184. C. J. Hilado and D. P. Brauer, Ignitability by radiant heat of furniture upholstery fabrics, *J. Consumer Product Flammability* **5**, 217–237 (1978).
185. C. J. Hilado and Debra L. Brandt, Ignition characteristics of some aircraft interior fabrics, *J. Fire Flammability* **9**, 527–533 (1978).
186. L. Benisek and W. A. Phillips, The importance and relevance of burning behavior, smoke and CO emission for upholstered seating, *J. Consumer Product Flammability* **5**, 96–110 (1978).

187. F. H. Praeger, Study of the burning behavior of upholstered furniture, *J. Fire Flammability* **9**, 71–87 (1978).
188. V. Babrauskas, Full-Scale Burning Behavior of Upholstered Chairs, TN 1103, National Bureau of Standards, Washington, D.C. (1979).
189. D. Harabin, E. R. Ostrander, and E. E. Stout, Properties of Textile Upholstery Fabrics and Their Importance to Consumer Satisfaction, Bulletin 1032, Cornell University Agricultural Experiment Station, Ithaca, New York (1973).
190. R. C. Maggio, A Molded Skin Simulant Material with Thermal and Optical Constants Approximating Those of Skin, Report NS 081-001, U.S. Naval Materials Laboratory (1956).
191. J. M. Davies, B. McQue, and T. B. Hoover, Heat transferred by decomposition products from cotton fabrics exposed to intense thermal radiation, *Text. Res. J.* **35**, 757–769 (1965).
192. E. Braun, V. B. Cobble, J. F. Krasny, and R. Peacock, Development of a Proposed Flammability Standard for Commercial Transport Flight Attendant Uniforms, Report No. FAA-RD-75-176, NTIS No. ADA-033-740, National Technical Information Service, Springfield, Virginia (1976).
193. R. M. Perkins, J. F. Krasny, E. Braun, and R. Peacock, An evaluation of fabrics for thermal protective clothing, in Proceedings of the Twelfth Annual Meeting, Information Council on Fabric Flammability, New York, pp. 212–230 (1978).
194. ASTM D 13-77-4, Proposed Test Method for Thermal Protective Clothing, American Society for Testing and Materials, Philadelphia (1977).
195. A. M. Stoll and M. A. Chianta, Method and rating system for evaluation of thermal protection, *Aerospace Med.* **40**, 1232–1238 (1969).
196. Specification for Protection Against Intense Heat for Short Periods, BS 3791, British Standards Institution, London (1970).
197. J. Quintiere, Radiative characteristics of firefighter's coat fabrics, *Fire Tech.* **10**, 153–161 (1974).
198. W. F. Baitinger, Product engineering of safety apparel fabrics: Insulation characteristics of fire retardant cottons, *Text. Res. J.* **49**, 221–225 (1979).
199. W. F. Baitinger, Fire retardance characteristics of safety apparel fabrics, in Proceedings of the 1978 Symposium on Textile Flammability, pp. 168–185, LeBlanc Research Corporation, East Greenwich, Rhode Island (1978).
200. R. H. Barker, personal communication (1979).
201. K. E. Perepelkin and B. A. Muchin, Der Einfluss der molekularen und supramolekularen Struktur der Chemiefasern auf ihre thermischen Eigenschaften, *Lenzinger Berichte*, pp. 46–66, Chemiefasern Lenzing A. G., Lenzing, Austria (1975).
202. Man-Made Fibers Fact Book, 12, Man-Made Fibers Producers Association, Inc., Washington, D.C. (1978).
203. R. C. Nametz, Flame retarding synthetic textile fibers, *Ind. Eng. Chem.* **62/3**, 41–46 (March 1970).
204. Polychals in Synthetic Fibers, Kohjin Co., Osaka (1968).
205. I. M. Gottlieb and L. R. Beck, Jr., Thermal stability, flammability, and melting characteristics of textile substrates, *Ann. N.Y. Acad. Sci.* **82**, 724–743 (1959).
206. L. Benisek, Die Entflammbarkeit von Fasern und die FlammResistenz der Wolle, *Textilveredlung* **8**, 318–320 (1973).
207. U. Einsele, Über das Brennverhalten und den Brennmechanismus von Synthesefasern, *Mell. Textilber.* **53**, 1395–1402 (1972).

5

The Analysis of Polymers and Polymer Degradation Products by Mass Spectrometry

Stanley C. Israel

1. Introduction

Since the development of the modern mass spectrometer slightly over 25 years ago, this technique has been applied to all branches of the chemical sciences. The utility of the mass spectrometer to identify and elucidate the structure of chemical compounds is so great that it is now considered one of the "classical" instrument techniques. The applications of mass spectrometry to the analysis of polymers and polymer degradation products offer the same unique advantages with respect to identification and structure determination as well as providing a rapid, extremely sensitive method for the characterization of polymeric materials individually and in mixtures. In the case of polymer degradation studies it allows for the analysis of the products resolved on a temperature or time basis. That is, in a study where the temperature is changing, analysis may be performed throughout the temperature range and the products and products concentration correlated with sample temperature. In the case of isothermal and pulse heating modes and degradation due to mechanical stress, a fast-scanning mass spectrometer can be employed to identify the products relative to their time of evolution.

Stanley C. Israel • Polymer Science Program, Department of Chemistry, University of Lowell, Lowell, Massachusetts 01854.

The effect of interaction between different components in a system undergoing degradation may also be efficiently studied. By this method we can distinguish between the additive and synergistic effects of additives to polymer systems and determine what species are produced by the interaction of the two or more substances. Thus the degradation mechanisms and mechanism of action of additives in polymeric systems may be fully elucidated. Other distinct advantages of mass spectrometric analysis are the very small samples needed (usually in the microgram to milligram range), and with the incorporation of a data system, analysis time may be reduced to a few minutes.

Basic Principles of Mass Spectrometry

The basic components of a mass spectrometer are shown schematically in Fig. 1. Operationally, a sample is introduced into the ion source, vaporized, and ionized. The ions which are produced are then accelerated out of the ion source into the mass analyzer, which is capable of discriminating among the gaseous ions based on the mass-to-charge ratio (m/e). There are a number of different types of mass analyzers in common use. The magnetic and electrostatic analyzer both operate on the principle that when a charged particle enters either a magnetic or electrostatic field, the field will deflect the particle through a radius dependent on field strength and momentum of the particle. The double-focusing mass spectrometer combines both a magnetic field and an electrostatic field in order to achieve a higher degree of discrimination among ions (high resolution). The quadrupole mass spectrometer makes use of four parallel rods which, upon application of a radio frequency and an applied voltage, set up an oscillating quadrupole field that will only pass ions of a specific mass. By varying the applied voltage, the mass being analyzed is varied. The only other mass analyzer that is commonly in use is the time-of-flight (TOF) analyzer, which functions by accelerating ions through a specific voltage drop. The velocity of each ion, accelerated in this fashion, is dependent upon its mass with the lighter ions traveling faster. The analyzer discriminates between ions by electronically relating time of flight between the source and the detector to the mass-to-charge ratio of the ion.

When the ions leave the mass analyzer they impinge on a detector. The detector must be sensitive enough to detect very few ions impinging on it per second, and capable of amplifying the signal to a level that will drive an output device. The most common detectors are the electrometer amplifier and the electron multiplier.

The modern mass spectrometer is based on the analysis of positive ions in the gaseous state. The process of producing these gaseous ions is critical in determining what materials may be analyzed by this technique. The most common method employed for the production of ions in the mass spectrom-

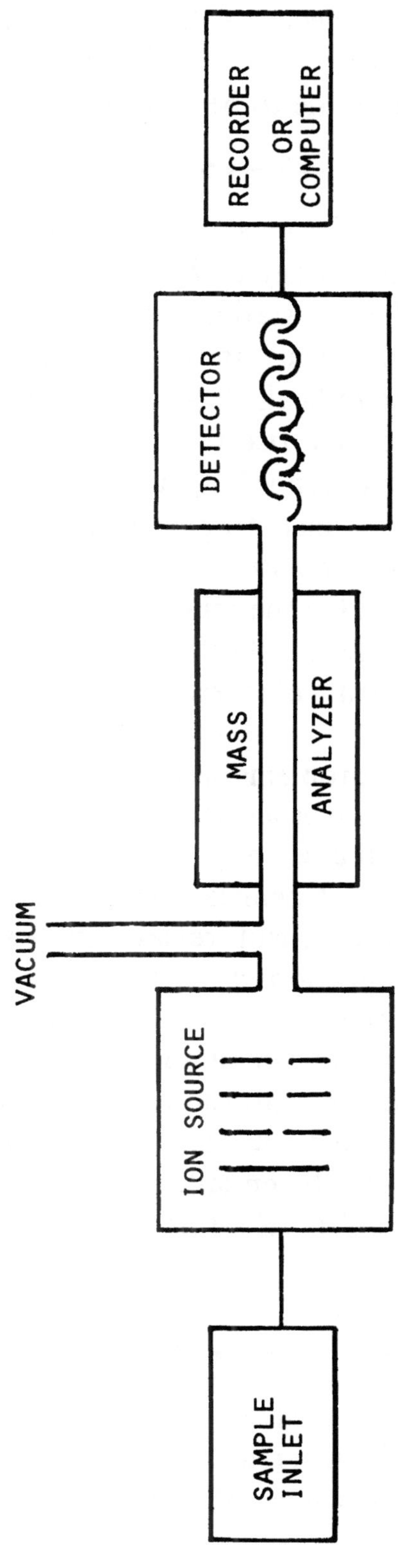

FIGURE 1. Schematic representation of a mass spectrometer.

eter is electron impact (EI), in which the gas phase sample molecules enter the ion source and are bombarded by a beam of electrons. If the energy of the electron is in excess of the ionization potential (more correctly, the appearance potential) the molecule will undergo ionization and form a positive radical ion (molecular ion) as represented by Eq. (1), where M represents the gas phase molecule to be analyzed.

$$M + e^- \longrightarrow M^{\ddagger} + 2e^- \quad (1)$$

Typical appearance potentials for most organic molecules fall in the range 7–13 eV. Routinely an electron energy of 70 eV is used and thus the molecule absorbs energy in excess of its ionization energy. This excess energy may go into breaking one or more chemical bonds causing the molecular ion to break down into fragment ions and radicals. It is these molecular ions and fragment ions that are analyzed and give rise to the classical mass spectrum. The fragmentation of the molecular ion is both an advantage and disadvantage of electron impact ionization. Since fragmentation is dependent on bond characteristic, the fragmentation pattern offers much information regarding the structure of the molecule. On the other hand, the fragmentation pattern renders the analysis of mixtures most difficult, as very little information can be gained from the mass spectrum of several compounds superimposed on one another.

In order to avoid the problems associated with fragmentation, it is necessary to use an ionization method that does not impart excess energy to the molecule. Ionization by the technique of chemical ionization (CI) is one method of achieving this aim. Chemical ionization was first developed by Munson and Field,[1] and excellent review articles have recently been published.[2,3] Chemical ionization is essentially a gas phase acid–base reaction in which the acidic species is formed by subjecting a reagent gas, e.g., methane, to bombardment by a beam of high-energy electrons. The methane is ionized [Eq. (2)] but the concentration of methane is so high that the methane radical cation will react with neutral methane molecules starting a series of ion molecule reactions. Essentially this process can be represented by the formation of the species CH_5^+ [Eq. (3)]:

$$CH_4 + e^- \longrightarrow CH_4^{\ddagger} + 2e \quad (2)$$

$$CH_4^{\ddagger} + CH_4 \longrightarrow CH_5^+ + CH_3 \quad (3)$$

The species CH_5^+ is a very strong acid and other organic molecules will be basic compared to it. Thus the CH_5^+ will react "chemically" in an acid–base type of reaction with any sample molecules (M) that it may encounter:

$$CH_5^+ + M \longrightarrow MH^+ + CH_4 \quad (4)$$

Thus the sample molecule has been ionized through the gain of a proton, forming a species represented by MH^+. This process can be controlled so that no excess energy is imparted to the molecular ion, and thus under ideal conditions no fragmentation will occur. The CI mass spectrum for a particular sample should give a single peak at a mass-to-charge ratio (m/e) one mass unit higher than the molecular weight of the compounds. Chemical ionization has the advantage that it will yield a species related to the molecular ions. Consequently the molecular weight for a compound that does not yield a molecular ion under electron impact may be obtained by chemical ionization. This facilitates the analysis of mixtures of compounds. However, the structural information that is inherent in the fragmentation pattern of EI spectra is lost.

Both electron impact and chemical ionization suffer from the limitation that the sample molecule must be in the gaseous state in order to be ionized. The techniques of field ionization (FI) and field desorption (FD) attempt to circumvent the volatility limitations of many substances. By taking advantage of the fact that it is the positive ions that are analyzed, it is only necessary to get the ions into the gaseous state and not the material itself. Inghram and Gomer[4] and Beckey[5] have found that molecules in very strong electric fields will become strongly polarized and will lose electrons to form ions. In field ionization mass spectrometry the electric field has a steep gradient and the polarized molecule will migrate to the positive end where an electron is stripped from it and the remaining positive ion is repelled out of the field and into the analzyer of the mass spectrometer.

The process of field desorption (FD) is essentially a modification of field ionization using a field anode that has been activated by benzonitrile or acetone vapor at temperatures in excess of 1200° C. The activation process leaves the anode covered with a large number of carbonaceous microneedles. The activated anode is either dipped into a saturated solution of the sample or the sample is dispensed onto the anode. The anode is then placed into the ion source and any solvent pumped off. The resulting solid sample is ionized in the electronic field and the resulting ions desorbed into the vapor state. As with chemical ionization, FI–FD techniques do not impart extra energy to the ion and thus fragmentation does not take place. Field ionization[6,7] and field desorption[8] have been reviewed in the literature.

2. *Direct Analysis of Polymers*

The analysis by electron impact mass spectroscopy requires that the sample have a certain degree of volatility at temperatures up to about 300°C and pressures to 10^{-6} torr. Very few polymeric systems meet these requirements and thus the direct analysis of most polymers is not possible. One polymeric system that is volatile under these conditions is poly(perfluoropropylene oxide). Ligon[9] found that low-molecular-weight fractions of poly(perfluoropropylene oxide) gave a significant number of evenly spaced ions of adequate and similar intensity in the mass range up to 2000 amu. Thus Ligon proposes this system as a mass spectral mass standard for the mass range 31–2000. At the present time he is investigating the extension of this study to the 2000–3700-amu mass range. Lee and Sedgewick investigated the homo- and copolymerization of ethylene oxide and propylene oxide. Within a limited molecular-weight range (DP $\leqslant$ 14) these polymers were volatile enough to be analyzed by mass spectrometry.[10] The mass spectra of the homopolymers showed a repetitive fragmentation pattern, with the pattern repeating every 44 amu for poly(ethylene glycol) and every 58 amu for poly(propylene glycol) corresponding to one "monomer" unit. Random and block copolymers were prepared and the mass spectra of these systems clearly indicated the polymer structure. In the case of the block copolymers the order in which the blocks were incorporated could easily be deduced. Considering that the intensity of a peak in the mass spectrum is proportional to its relative abundance, Lee and Sedgewick determined the molecular weight distribution for each of the polymers. They calculated the average molecular weight for each polymer and in the case of the block copolymers calculated the average molecular weight of the constituent blocks. The values of the average molecular weight determined mass spectrometrically were in good agreement with the value determined by end-group analysis. The molecular weight distributions were used as the basis for a kinetic analysis of the polymerization reactions.[11] The calculated rate constants decrease rapidly for the formation of the first three or four units followed by less of a decrease at higher units, indicating a complex reaction scheme in which the rate of reaction of the growing chain becomes slower as the chain length increases.

The gamma-ray-induced polymerization of 1-vinyl-1,2-dicarbo-*closo*-dodecarborane was investigated by Wright and Klingen.[12] This monomer, which exists in a type of mesomorphic state at room temperature, was polymerized in high yield with the formation of no detectable side products. In order to determine the basic structural unit the polymer was subjected to mass spectrometric analysis. The EI spectra of dimer, trimer, and tetramer indicated a linear "head-to-tail" type of structure with each polymer chain containing a terminal vinyl group. A similar approach was taken to elucidate the structure and steriochemistry of a series of oligormeric poly(*n*-alkylimino-alanes).[13] The mass spectrometric, X-ray diffraction, ^{1}H, and ^{13}C NMR data led the authors to conclude that these compounds formed three-dimensional oligomers by association of the monomer units via four- or six-membered rings.

Lattimer *et al.*[14] have investigated the field desorption mass spectrum of a series of low-molecular-weight polymers which were of low volatility and thermally labile. Thus analysis by either electron impact or chemical ionization was not possible. The polymers that were investigated included poly(trimethyldihydroquinoline) (poly-TMDQ), polystyrene, and poly(propylene glycol). In the case of poly-TMDQ and polystyrene, the molecular ions arising from the oligomers were observed up to a mass of 1500 amu. In the case of polystyrene the mass spectrum indicated a butyl-terminated polymer. The FD mass spectrum of the poly(propylene glycol) shows mostly protonated molecular ions for the oligomers to mass 947 (DP = 14). The ions were "offset" by 18 amu, indicating hydroxyl termination.

The further development of field desorption techniques is an active area of research at the present time. As these newer techniques are applied to polymers, the number and variety of systems analyzed directly should greatly increase. However, as the molecular weights of the polymers increase, it becomes increasingly more difficult to desorb or vaporize even an ionized molecule, and the mass spectrometer itself has inherent limitations that work against the analysis of very-high-molecular-weight molecules. In order to analyze these very-high-molecular-weight species, instruments with specific design criteria for this purpose will have to be built. Thus it is expected that we soon will reach an upper limit on the size of polymer molecules that can be analyzed directly. It is doubtful that this upper limit will exceed a molecular weight of 100,000.

3. *Analysis of Polymer Degradation Products*

Polymers are subject to degradation by a great variety of forces. Thermal, mechanical, and electrical stress, as well as radiation, will cause bond cleavage in polymeric systems. The understanding of the nature of these processes and the kinetics of these reactions have commanded much scientific attention. Much of the interest centers on determining the mechanism of the degradation process and the effects that modifications and additives have on degradation. Since the degradation process produces smaller and generally more volatile fragments, the mass spectrometer has great utility in determining the identity and concentration of the degradation products, and also, to follow the degradation process itself by monitoring the evolution of those products.

3.1. *Thermal Degradation*

Thermal degradation, followed by analysis of the degradation products, has been a standard method of establishing polymer structure.[15-18] Knowledge of the rate of degradation and identity of the degradation products is essential in determining the mechanism of degradation, as well as to establish the effects of polymer modification on the mechanism.

The process of thermal degradation can be divided into two types: pyrolysis and combustion. Pyrolysis can be defined as thermal degradation in the absence of oxygen, and combustion can be defined as thermal degradation in the presence of oxygen. Combustion can be studied in the flaming and nonflaming mode. Both of these processes can be carried out under a variety of conditions, in a number of different types of apparatus. The most common techniques of thermal analysis can be classified based on the rate of temperature rise or temperature rise time (TRT).

Isothermal and slow heating rate techniques usually employ heating rates up to about 100°C/min. This condition is achieved by a wide variety of equipment ranging from the heating of the sample in a flask over a flame, to ovens and tube furnaces, to very sophisticated thermobalances and microreactors.[19,20] The mode of heat transfer may be by direct contact or may be radiative, or a combination of both, and depending on the particular apparatus, degradation may routinely be conducted in oxygen, inert atmospheres, or under vacuum.

Faster heating rates (up to about 100°C/msec) are achieved with any of three basic probe-type units. The filament type uses a platinum or gold ribbon or wire filament that is resistively heated.[21,22] A second type incorporates a controlled capacitor discharge circuit to rapidly raise the temperature of the filament to its final value,[19,21,23] and the third type, known as the Curie-point pyrolyzer, is inductively heated to the Curie point of the filament metal.[24,25] In all probe-type units the sample is normally placed directly on the filament and the heat transfer is by direct contact. The use of metals in filament and Curie-point apparatus gives rise to the possibility of catalytic effects by the metal on the degradation, which, in fact, are reported to occur with different metals yielding different degradation products and product distributions.[26] The systems may also be designed with the sample isolated from the filament by a sample holder (normally a quartz tube) and the sample subjected to radiant heat transfer. The probe-type units may be configured to operate under oxidizing and nonoxidizing atmospheres, but the majority of work to date has been performed under vacuum or inert gas conditions.

Ultimate heating rates (up to and exceeding 100°C/μsec) can be achieved by laser microprobe techniques. A laser is a unique tool that will focus an intense beam of precisely known energy on a sample. The laser typically can be pulsed with a pulse width on the order of 0.001 sec producing incident energies on the order of 5 J/pulse. If this energy were to be focused on a spot of 0.1 cm diameter the power deposited on the sample would be 6.4×10^5 W/cm^2.[27] The final temperature attained under these conditions is very difficult to determine. Vanderborgh estimates temperatures in excess of 10^5 K as measured by the velocities of the atomic fragments.[28] The final temperature is reached within the time frame of the pulse width, thus yielding temperature rise times of approximately 10^8 K/sec. This is in agreement with Fanter and

Levy,[29] who estimate temperature rise times of approximately 10 msec to reach equilibrium temperatures in the range 927–1227°C. An inherent problem in laser pyrolysis is the fact that many samples are transparent to the laser frequency. The ruby laser was the first of the high-energy output lasers available. It emits at a wavelength of 694.3 nm. Most organic materials are transparent to this wavelength, making it necessary to mix an energy absorbing material, for example, graphite,[30] with the sample to effect energy absorption and heat transfer.[31] In order to achieve efficient heat transfer into the sample, the graphite and the sample must be finely powdered. This gives rise to a large amount of absorbed gases which are very difficult to remove even under high-temperature and high-vacuum conditions.[32] In fact, in their early work in laser pyrolysis–mass spectrometry, Joy and Reuben[33] found that they could not distinguish between the polymer decomposition product and the gases that were absorbed on the surface of the materials. It has also been observed that the addition of graphite can alter the pyrolysis product distribution.[34] Vanderborgh has reported on the successful use of powdered nickel mixed with the sample as an energy-absorbing media. He observed no catalytic effects due to the nickel on the pyrolysis pattern.[35] Walker[36] and Fanter *et al.*[29] have achieved pyrolysis by coating a thin layer of sample on cobalt glass sample holders to absorb the laser energy and transfer it to the sample. With the development of lasers emitting at wavelengths lying within the vibration bands of organic materials, it is possible to have direct absorption of energy by the material.[37] At present only the CO_2 laser with spectral output in the infrared range and the argon-ion laser with spectral output in the visible range are in common use. Marked differences are noted when laser pyrolysis patterns are compared to those produced by other means.[27] The number of compounds produced appears to be far less, and of higher molecular weight under conditions of laser-induced pyrolysis, compared to filament-type pyrolyzers. Vanderborgh speculates that laser-induced pyrolysis is primarily a shock phenomenon and the higher-molecular-weight fragments presumably result from thermal blow off and are a simple and predictable group.[27]

Flaming combustion is the one degradation technique that cannot be described by a temperature rise time. In practice, the rate of temperature rise for a flaming combustion probably cuts across the complete range of TRT with heat being transmitted by all modes of conduction and radiation. The degradation processes taking place with the flaming system are a complex mixture of pyrolysis, nonflaming combustion, and flaming combustion. The criteria for any degradation process is reproducibility. In order to achieve this in flaming combustion, steady state combustion must be established. On a laboratory scale this condition is very difficult to attain, especially when dealing with the burning of materials, as opposed to pure fuel gases. A small flame is inherently unstable and the failure to control it leads to irreproducibility, as indicated by the varying combustion product distributions observed. The source of much of the flame instability is the ignition process which, by definition, is not a steady state process. The ignition of the sample and the propagation of the flame is often a violent process, accompanied by strong pressure and temperature gradients and is extremely sensitive to environmental effects and disturbances.

The different degradation techniques give rise to differences in degradation products and to product distributions. The isothermal and slow heating rate techniques usually require that the degradation products reside in, or pass through, the heated zone for a significant amount of time, and thus the degradation products are subjected to secondary thermal degradation. This leads to a dependence of the final product distribution on the resonance time of the gases in the heated zone.[38,39] In probe or laser techniques the pyrolysis products move away from the heat source, resulting in little opportunity for secondary degradation.

Flaming combustion incorporates all of the problems of all the other techniques. In a flaming combustion system all three thermal degradation modes (pyrolysis, nonflaming combustion, and flaming combustion) are occurring simultaneously in different regions of the system and degradation products from different parts of the sample spend varying amounts of time in the heated zone.

Each technique of thermal degradation yields a mixture of products, and it is this mixture of products that the mass spectrometer must be able to analyze. There are three approaches to the mas spectrometric analysis of this mixture of degradation products. The first approach is to interface the degradation system directly to the mass spectrometer and analyze the products directly. The second approach is to collect the degradation mixture and subject it to a separation procedure before analysis. The third approach is based on the philosophy that the pyrolysis of a material will yield a unique mix of products characteristic of that material. Thus the spectra taken on the degradation mixture resulting from the pyrolysis of a material should be unique and provide a "fingerprint" of the material. This technique is not utilized in an attempt to identify any of the individual degradation products.

3.1.1. *Direct Analysis of Thermal Degradation Products*

By the first approach, the degradation system is interfaced directly to the mass spectrometer, thus the degradation products enter the ion source of the mass spectrometer as they are evolved from the material. Since the products are analyzed immediately, the course of the degradation may be followed directly. In the case where the degradation is conducted isothermally, it is possible to monitor the evolution of the degradation products with time. If a temperature-programmed heating mode is employed, the analysis yields a relationship between the evolved products and the temperature of degradation. Not all degradation modes lend themselves to interfacing directly to the mass spectrometer. The ion source normally operates at a pressure of about 10^{-6} torr, and unless special interfaces are used, only degradation under vacuum is possible.

The interfacing of a mass spectrometer to a controlled temperature cell was first described by Langer and Gohlke[40] and was applied to polymers by Friedman[41,42] and Shulman.[43] By this technique the sample is heated either isothermally or at a controlled temperature rate. The pyrolysis gases are continually analyzed by the mass spectrometer. By recording the total ion signal vs. temperature, a mass thermal analysis (MTA) curve may be generated. The great advantages of this technique over a conventional thermobalance analysis is the ability of the mass spectrometer to identify and follow the evolution of each degradation product. Where the thermobalance only determines weight loss and thus cannot resolve competing or overlapping reaction, the mass spectrometer will allow the generation of a thermal analysis curve for each degradation product, thereby resolving competing and overlapping reactions. This, in principle, permits a complete analysis of decomposition kinetics and mechanisms. Friedman[42] demonstrated the utility of this technique by examining a series of polybenzimidazoles. The degradation of these polymers produced complex TGA curves. The mass spectrometric analysis was able to identify about 15 different degradation products and individual mass thermal analysis curves or mass fragmentograms (ion signal vs. temperature) for the major decomposition products were plotted. In contrast to the TGA curves, the MTA curves were smooth and well resolved.

When a mixture of compounds is introduced into the mass spectrometer the result is a composite spectra which is made up of the individual spectra superimposed on one another. The interpretation of this composite spectra is often extremely difficult except in the case where one or more of the components in the mixture has a characteristic peak in the mass spectra that is lacking in the spectra of all the other components. This approach is exemplified by a study of Jacobi and Luderwald[44] in which the direct pyrolysis–mass spectrometry of homo- and copolyesters of glycolide and L-lactide of known compositions were investigated. They found that the fragment at m/e 42 was specific for lactide units and the one at m/e 116 was specific for glycolide units. He generated a standard curve that allowed for the determination of relative amounts of each ester unit in copolyesters of unknown composition.

Mol[45] assembled a directly coupled Mettler thermoanalyzer and quadrupole mass spectrometer. The quartz thermoanalyzer furnace was modified to accept the mass spectrometer in a vertical position directly above the sample pan. The distance between the sample pan and mass spectrometer was kept at a minimum, only 20 cm. The furnace and spectrometer were fitted with a pneumatic lifting mechanism which allowed sample changing without the necessity of breaking the furnace to spectrometer seal. During a typical analysis the thermal analyzer was heated at a programmed rate and the mass spectrometer continuously scanned. The sample weight loss, total pressure, and mass spectrometric data were plotted as a function of temperature. From the simultaneously generated TGA, DTA, DTG, pressure, and MTA plots the kinetics and decomposition mechanism was determined and the activation energy and order of reaction calculated for the overall reaction and for the formation of specific species identified in the decomposition products. Mol reported on the decomposition of TDI- and MDI-based polyurethanes, a capped and uncapped polyester, and an ABS plastic. In the case of the ABS plastic, from the mass spectra it was possible to follow the polyacrylonitrile degradation by the appearance of HCN. Butadiene is produced from the polybutadiene degradation and the decomposition of polystyrene yields styrene monomer. The MTA curves for HCN, butadiene, and styrene are unique and may be used for classifying different types of ABS. In this degradation the spectra taken at 120°C indicated the evolution of low-molecular-weight hydrocarbon. At about 260°C the appearance of HCl becomes evident and its evolution reaches a maximum at 284°C. The HCl is attributed to the dehydrohalogenation of PVC employed in the ABS formulation as a fire retardant. By 400°C the mass peaks characteristic of the decomposition of the acrylonitrile, butadiene, and styrene becomes evident and the decomposition was almost complete by 468°C. The MTA curves were very well resolved and were shown to be far more sensitive to the detection of additives and other volatiles than conventional thermogravimetry.

Carroll-Porezynski[(46)] reported on the utility of a DTA–MS analysis for predicting the flammability of composite material. They evaluated the compatability, thermal behavior of evolved gases of the individual components, and of the composite fabrics and polymer and oxidative degradation, in an attempt to elucidate the synergistic effects and combustion mechanisms. TGA–MS has been used to advantage in the study of a number of polymeric systems including PVC,[(47)] PMMA,[(48)] polyurethanes,[(47)] polystyrene,[(49)] and quinoxaline derived polymers.[(50)] Korobeinichev[(49)] reported on the effect of NH_4ClO_4 on the thermal degradation of polystyrene.

In the case of the thermal degradation of PVC, Mickelson[(51)] attempted to determine whether the dehydrochlorination step (formation of HCl) precedes the chain scission step (formation of benzene) or do the two processes occur simultaneously. By interfacing the mass spectrometer directly to the TGA and continuously monitoring $m/e = 78$ (benzene) and $m/e = 36$ (HCl) during the decomposition of PVC under helium, it was observed that simultaneous evolution of HCl and benzene occurred with a mole ratio of 2.5:1, respectively. These observations suggest that at least at these temperatures both dehydrochlorination and chain scission occur simultaneously.

Much of the effort in this area has been devoted to overcoming the numerous instrumental problems encountered in interfacing the two techniques. The inherent problem in interfacing a thermobalance or any other thermal device to a mass spectrometer is the necessity of transfer lines between the two. The transfer lines must be heated to prevent condensation and thus the degradation products may undergo secondary degradation in the transfer lines. Depending on the material of construction, the transfer lines may give rise to catalytic effects on the degradation products. This problem may be avoided by performing the thermal analysis directly in the ion source of the mass spectrometer. One way this may be accomplished is to place the sample in a temperature-controlled direct insertion probe. The temperature may be programmed at a controlled rate and the mass spectrometer can periodically scan the full spectrum, or alternatively, it can be set to monitor a single mass or a number of specific masses of interest. This approach was first suggested by Zitomer[(52)] and some early work in this area is described by Murdoch and Rigby.[(53)]

Ryska[(54)] used a temperature-programmed thin platinum wire probe in the ion source thermal analysis of poly(vinylacetate), poly(vinylchloride), poly(methyl methacrylate), and polystyrene. The MTA graphs were quite reproducible and readily distinguished between vaporization of solvents and unreacted monomer and pyrolysis products. Wiley conducted a series of studies reporting on the mass spectral characteristics, including metastable transitions and thermal degradation mechanisms for a series of polymers in the ion source of a cycloidal path mass spectrometer. The polymers studied included cross-linked polystyrene,[(55)] polyxylylene,[(56)] polypivalolactone,[(57)] and poly(2,6-dimethyl-1,4-phenylene ether) (PDPE).[(58)] Luderwald and coworkers have employed this technique in a number of systems. Using a gold wire heating probe they have studied aromatic and aliphatic polyesters,[(59)] polysulfonates,[(60)] polyamides,[(61,62)] and copolyamides.[(61–63)] This technique has found acceptance as a method of characterization of polymeric materials for quality assurance and control. For this purpose it has the advantage of causing less breakdown than a rapid pyrolysis and the fragments are larger and more characteristic of the starting materials. Nylons,[(64)] polyurethane foams,[(64,65)] silicone rubbers,[(64)] and phenylformaldehyde resins[(64)] have been characterized by this method. In the case of polyurethane foams, it is possible to identify the blowing agent and isocyanate used and to characterize the polymer type as a polyester or a polyether.[(64,65)] The advantages of this technique are that it is reasonably quick and easy to use and it requires a minimum of sample preparation and only minute amounts of sample.[(64,65)]

Stenzenberger[(66)] applied the techniques of pyrolysis–field ionization mass spectrometry and TGA to the thermal degradation of poly(bismaleimides). He conducted a classical thermogravimetric study generating temperature vs. weight loss curves (TGA) and differential (DTG) curves. In order to determine the chemical identity of the decomposition products, separate pyrolysis experiments were performed and the gaseous products analyzed by field ionization mass spectrometry. The samples were pyrolyzed in a gold or quartz vessel which was brought to the pyrolysis temperature by an electrically heated Pt coil. The degradation was run in a vacuum and the pyrolyzate collected in the evacuated gas inlet system of the mass spectrometer. The FI–mass spectrum was one of predominately molecular ions which could be related directly to the fragments produced by the thermal degradation. Lum applied the technique of laser-induced heating to the mechanism of polymer decomposition.[(67,68)] He used a focused argon-ion laser to irradiate the samples at low power densities. Detection of the degradation products was accomplished by a modulated molecular beam mass spectrometer. In modulated beam mass spectrometry the degradation products pass into the ion source as a molecular beam. This beam is modulated by a mechanical chopper, and the analysis accomplished by a quadrupole mass filter. The modulated beam allows for the electronic separation and subtraction of the spectra of the residual gas in the background from the spectra of the molecular beam. Thus only the spectra arising from the sample were recorded. Lum[(67,68)] applied this technique to the elucidation of the mechanism of PVC pyrolysis. In the degradation of pure PVC the major peaks observed were due to HCl^{35} and HCl^{37} (m/e 35–38), benzene (m/e 50–52 and 73–79), and a small amount of toluene (m/e 91–92). Higher-molecular-weight species were observed but at very low intensities. The peaks due to these three species were free from interferences and the intensity of the signal at these masses was directly proportional to the amount of that species produced in the degradation. By continuously monitoring the signal at the particular mass of interest, Lum was able to follow the evolution of HCl, benzene, and toluene with time. He found that a small amount of HCl was released first at low temperatures followed by simultaneous release of all three species as the temperature was raised. At faster heating rates the initial release of HCl was not observed and all three species reached their maximum evolution rate at the same time. Upon addition of the plasticizer di(2-ethylhexylphthalate) (DOP) to the PVC it was found that the evolution of HCl and benzene followed the same profile as in pure PVC but the toluene evolution profile was radically changed. Lum extended this work to a very elegant study of the synergistic effects between PVC and antimony oxide.[(67–69)] The laser-induced pyrolysis of a PVC–antimony oxide formulation yielded the first direct observation of the production of $SbCl_3$ during thermal degradation. There was no evidence of volatile SbOCl or any other antimony compounds in the spectrum. In order to elucidate the nature of the $SbCl_3$ formation a sample of pure PVC and a sample of pure antimony oxide were placed in the pyrolysis cell such that there was no physical contact between the two samples. The laser beam could be focused to impinge on only one of the samples exclusively. When the antimony oxide was selectively irradiated in the presence of unheated PVC there was no evolution of any detectable species. However, when the PVC was selectively irradiated in the presence of unheated antimony oxide $SbCl_3$ was observed in addition to the normal PVC degradation products. Thus it was concluded that the $SbCl_3$ was formed as a direct result of a chemical reaction between antimony oxide and the HCl evolved from the PVC and the reaction proceeds in the absence of any thermal decomposition of the antimony oxide. The chemical nature of this reaction was further confirmed by introduction of pure HCl into the sample cell containing only antimony oxide at room temperature. $SbCl_3$ and H_2O were observed in the spectra of the molecular beam emanating from the cell. The same species were observed when pure HCl was added to the room temperature sample cell containing a sample of plasticized PVC containing antimony oxide. Thus Lum established the existence of a reaction pathway for the production of $SbCl_3$ from PVC formulations containing antimony oxide which does not involve the thermal degradation of any antimony compounds. The same technique was employed in the investigation of polyoxymethylene copolymer[68] and polyether ester copolymer.[68]

Another technique utilizing thermal degradation directly in the mass spectrometer makes use of a high-energy source focused on a thermally thick sample to degrade the sample. The energy impinging on the sample remains constant and the evolution of degradation products is measured with time. Thompson[70] demonstrated the utility of this approach in a study of polyesters. He used a laser microprobe to determine the structure of a series of polyesters and to determine the identity and distribution of additives and impurities through a cross section of the sample. The sample was irradiated in the ion source of a time-of-flight (TOF) mass spectrometer by focusing a Nd^+ glass laser through a low-powered microscope. This "microprobe device" pyrolyzed the sample beginning at the surface and eroding into the interior. The TOF mass spectrometer may be scanned very rapidly and a series of mass spectra are obtained through the cross section of the sample. By this method Thompson was able to qualitatively analyze the polyester for impurities, detecting Sb residues. The technique also allows for the analysis of short-range order in polymer systems. The spectrum showed significant differences between "normal" non-cross-linked and cross-linked polyester. They identified a number of peaks unique to the cross-linked material. Spectra taken of various gels, ranging from slightly to heavily cross-linked, all exhibited these unique peaks with their intensity roughly correlating with the amount of cross-linking. Thompson demonstrated the applicability of this technique to the analysis of surface contamination of metals. A microscope black deposit on the surface of an aluminum plate was analyzed and identified as polyisoprene.

These techniques have long been used in the study of metals and for the analysis of surfaces. Energy sources such as high-energy ion bombardment by sputter ion gun [ion scattering spectrometry (ISS)–secondary ion mass spectrometry (SIMS)] or high-energy sparks (spark source mass spectrometry) have long been used in this field. The technique of ion probe microanalysis–secondary ion mass spectrometry (SIMS) was described by Sacher(71) in the determination of the effect of outdoor exposure on polymers. They determined the finish thickness of an epoxy resin system on the surface of a glass–epoxy composite material. Measured amounts of the finish were sputtered off and introduced into the MS for analysis. The depth at which silicon-containing peaks were identified by the MS was interpreted as the thickness of the outer resin. This approach is beginning to be applied to various polymer systems.(72–79)

Fristrom(80–83) has taken a very novel approach to the study of flaming combustion of polymers. The basic concept of his approach is to exchange a transient time dependence for a steady state space dependence by moving the sample at a constant speed through a stationary ignition source. The technique known as the moving wire technique (MWT), requires the polymer to be coated on a wire and moved at a speed through a tent-shaped flame produced over a slot burner. The system is at equilibrium and allows the precise fixing of the location where ablation, ignition, and extinction are taking place.

The analysis of the degradation products from the different locations are made through a quartz microprobe connected directly to a TOF mass spectrometer.(84) The microprobe was mounted on a three-dimensional micrometer and could sample from any location of interest. By this technique they have analyzed commercial samples of Teflon,(82) PVC,(82,83) and a rubber insulation.(82) In the degradation of PVC they identified ignition as occurring at the first appearance of a flammable gas identified as benzene. Samples of pure PVC gave off HCl prior to ignition but commercial PVC did not. The authors suggest a wide variety of conditions that may be modeled by this technique including simulation of pilot ignition, both in oxidizing flame gases and interfaces between a heat source and adjacent atmosphere stream.(83)

Direct analysis of degradation products is a very powerful technique for following the course of the degradation reaction directly. It suffers from the limitation that it is primarily applicable to degradation under vacuum. The number of products that may be identified by this method is also limited to those that have unique mass spectral features, and can be distinguished within a mixture. With the more widespread use of FI–FD and CI technique the ability to identify and monitor more of the products directly will be greatly increased.

3.1.2. Off-Line Analysis of Degradation Products

The second approach to the analysis of degradation mixtures is to use the mass spectrometer as an off-line device. Under these conditions the products are trapped or otherwise collected as they evolve from the sample. Many different type traps are available and they are quite adaptable to almost any type of degradation apparatus including the sampling from full-scale fires. At the end of the trapping period the trap is removed and the contents subject to analysis. When the mass spectrometer is used for off-line analysis it is possible to perform a very complete quantitative and qualitative analysis of the degradation products by first subjecting the trapped degradation products to a separation step prior to mass spectrometric analysis. The fewer the number of components in the mixture, the simpler the spectra, and thus the easier the identification. There are many analytical techniques available for the separation of volatile products. Among the ones most commonly applied to polymer degradation studies are differential condensation, gas chromatography, and high-pressure liquid chromatography.

Differential condensation is a trapping method that separates volatile materials on the basis of vapor pressure. The system consists of a number of cold traps, each being maintained at a different temperature connected in series. Typically about four or five traps are employed, maintained at temperatures of 0, −45, −75, −100, and −196°C. The volatile products pass first through the 0°C trap and all materials condensable at this temperature are collected. The remainder of the vapor then passes into the −40°C trap and those materials condensable at that temperature are collected. The gases move from trap to trap until all the condensables are trapped. At the completion of the experiment the contents of each trap are subjected to analysis. This fractionation method has found utility in conjunction with the technique of thermal volatilization analysis (TVA). The TVA technique is a method useful in the study of the characterization and thermal degradation process of polymers. The technique was first described in 1966 by McNeill.[85] A good description of the technique and experimental variables can be found in subsequent papers by McNeill.[86–88] TVA involves the controlled heating of a sample in a sample tube under a continuous pumped vacuum. The pressure of the system is continuously monitored by a Pirani gauge. When decomposition of the samples takes place the volatiles produced give rise to an increase in pressure in the system which is recorded by the Pirani gauge. The pressure rises then falls as the volatiles are pumped away and/or collected in the cold trap. The pressure, as measured by the millivolt output of the Pirani gauge, is recorded as a function of temperature. Depending on the mechanism of degradation, one or more peaks may be observed. In their study of the influence of ammonium phosphate on the thermal degradation of poly-(methyl methacrylate) Camino *et al.*[89] used the mass spectrometer to identify the products that were noncondensable at −196°C. These were identified as principally carbon monoxide with traces of methane.

Jamieson and McNeill studied the degradation of mixtures of poly(vinyl-acetate) and poly(methyl methacrylate)[90,91] and the effect of silver acetate[91] on the PMMA degradation. They used a twin-limbed degradation tube which enabled them to study the two polymers in an intimate mixture "mixed" or the two polymers simultaneously degraded in separate tubes "unmixed." They used four traps at 0, −75, −100, and −196°C, respectively, and followed the pressure after each one. The condensables were fractionated into those condensable at −80°C and those products noncondensable at −80°C but condensable in liquid nitrogen at −196°C. These products were analyzed by low-voltage EI mass spectrometry and infrared spectroscopy. The results indicated an interaction between the PVA and PMMA in the "mixed" state that is not observed when the degradation of the PVA and PMMA takes place simultaneously but separately and only the degradation products are allowed to mix ("unmixed" state).

In the analysis of the thermal degradation of polystyrene and poly(*p*-*N*,*N*-dimethylaminostyrene) (PDAS), Still[92,93] analyzed the fraction volatile at −80°C by direct introduction into the mass spectrometer. Methane, ethylene, and propylene were identified from PDAS, whereas polystyrene showed only ethylene in significant amounts.[92] The pyrolysis at temperatures between 340 and 500°C produced a viscous oil which lined the walls of the pyrolysis tube.[93] The oil was analyzed by placing it on the direct probe of the mass spectrometer and slowly raising the temperature from room temperature to 360°C. The oil fraction was effectively fractionated and at ambient probe temperature, the monomer was identified. As the probe temperature was raised above 120°C a number of different dimeric species were identified. At higher temperatures (between 250 and 360°C) saturated and unsaturated trimers and tetramers were apparent. The oil fraction for both PDAS and polystyrene gave essentially similar components as identified by mass spectrometry.

Chiu and Beattle[94] extended the technique of differential trapping to identify and quantify small amounts of comonomer in methyl methacrylate copolymers. They employed a four-trap system, three in series and one in parallel. The one trap in parallel is normally isolated from the system and is opened to the system to trap closely spaced components without interrupting the heating program. At the completion of the degradation the contents of each trap were analyzed by mass spectrometry. Almost 95% of the volatile products were trapped at dry ice–acetone temperature and these were identified as mainly methyl methacrylate monomer. The remaining products were trapped at liquid nitrogen temperature and it was the analysis of the contents of this trap that showed the differences between samples. Ethyl acrylate, methyl acrylate, methyl isobutyrate, and methyl propionate were identified in concentration ranging from 0.1 to 29 mol.% for the different samples.

In the study of thermal degradation of blends of polystyrene with polybutadiene, McNeill employed the technique of subambient TVA,[88,95] or, as it is also known, differential distillation,[96] for the analysis of products. By this technique the content of each of the cold traps in the TVA experiment is analyzed by slowly warming the trap to ambient temperature and analyzing each component as it distills from the trap. In the analysis of polystyrene and butadiene mixtures, McNeill found that as the warming rate was decreased, better fractionation was achieved and ultimately a large number of components were resolved and identified in the degradation mixture.

Gas chromatography (GC) is at present the ultimate and most widespread technique for the resolution of a volatile mixture into its individual components. The separation occurs in the chromatographic column containing a sorbent layer known as the stationary phase. A gas phase sample is introduced onto the head of the column and the sample is moved through the column by the flow of a carrier gas. As the sample or mobile phase moves through the column the individual components of the mixture partition themselves between the stationary and mobile phases. A dynamic equilibrium is established for each component between the two phases with each component having a different partition coefficient. Thus each component moves through the column at a different speed dependent upon its affinity for the stationary phase (partition coefficient). Under these conditions each component of the mixture exits the column at at a different time. With the development of fast-scanning mass spectrometers it was recognized that the interfacing of the two techniques would produce a gas chromatograph–mass spectrometer (GC–MS) system that could take a complex mixture, separate it into its individual components, and introduce those components individually into the mass spectrometer for analysis.

Samples may be introduced onto the column in many ways. For example, gases may be introduced directly onto the column by a gas-sampling loop. A mixture of volatile liquids may be injected into a heated injection port where the sample is vaporized and the vapors enter the column. A method that is widely used in polymer degradation studies is to trap the gases in a low-temperature or absorbent trap then transfer the trap to the injection port of the gas chromatograph and upon heating the contents of the trap are desorbed onto the column.

Gas chromatographic techniques are probably the most widely used for the identification and quantitation of mixtures of volatile liquids and gases. The use of GC–MS has found great utility in the analysis of impurities in monomer and polymer systems,[(64)] the study of the kinetics and mechanisms of polymer reactions, and, of course, for the analysis of degradation products. Thermal degradation studies employing GC–MS technique cut across all of the available thermal degradation methods, including flaming combustion.[(97–99)]

In the case where a polymerization reaction gives off volatile products, i.e., condensation polymerization, or where information on the rate of formation of oligomers is desired, the GC–MS may be used to follow the kinetics of the reaction. Linder[100] studied the reactions between phenol and formaldehyde. The reaction products were not volatile and thus they were silyated to render them volatile so that they could be analyzed by GC–MS. The liquid samples were injected directly into the heated injection port of the chromatograph. The analysis yielded over 20 peaks with the first ten being positively identified.

Sykes and Young[101] followed the degree of cyclization occurring in a series of four pyrrone and two polyimide prepolymers during curing. The materials were heated from 25 to 400°C at 2°C/min under helium. The effluent gas was sampled periodically and analyzed by GC–MS. The amount of H_2O and CO_2 evolved during the thermal cyclization indicated that the pyrrone polymers were less than 80% converted to the cyclized ladder polymers while the polyimides were fully converted to the cyclized structures under these conditions. The evolution of CO_2 by the pyrrone systems was given as the reason for the low conversion to ladder polymer.

Probe-type and laser pyrolysis may be conducted directly in the injection port of the gas chromatograph. The normal procedure is to conduct the pyrolysis in the carrier gas stream (usually helium) so that the flow carries the pyrolysis products onto the chromatograph column. A wide variety of systems have been studied using this approach.

Einhorn and co-workers have used probe pyrolysis extensively in the study of the thermal degradations of urethane foams.[102–105] Pyrolysis gas chromatograph–electron impact mass spectrometry was used to identify the volatile low-molecular-weight components. The less volatile higher-molecular-weight species, mostly the polyol fragments, were identified by gas chromatography combined with chemical ionization mass spectrometry. The authors have proposed a detailed thermolysis mechanism for these systems. Gritter *et al.* studied the decomposition of homopolymer and copolymer systems including polysulfones,[106] methacrylate polymers,[107] and copolymers of methyl methacrylate and methacrylic acid.[107] The thermal degradation was carried out by probe pyrolysis and the products analyzed by GC–MS. In the case of PMMA they confirmed the results of previous work, obtaining predominantly monomer upon pyrolysis, but they also identified small amounts of other products including saturated ester and methyl isobutyrate, which they argue is formed by free-radical reaction. In comparison to PMMA, the PMAA homo- and copolymers yielded a large number of identifiable degradation products including both unsaturated aliphatic and aromatic hydrocarbons. The extensive decomposition of the MAA during pyrolysis indicates that the chemistry of this decomposition is considerably more involved than the mechanism proposed by Grassie.[108] Gritter also observed that the copolymers were readily contaminated by metal ions when polymer work-up included a precipitation into tap water, and the inclusion of these ions had a direct affect on the pyrolysis products. Probe-type pyrolyzers have been combined with the GC–MS to study a number of polymeric systems including PVC,[109,110] polystyrene,[111] polymeric sulfanilamides,[112] formaldehyde polymers,[113] fluoropolymers,[114–116] and perfluoroalkylene-linked organo siloxane polymers,[115] poly(propiolactones),[117] vinyl acetate,[118] polyethylene,[109] vinyl acetate–olefin copolymers,[118] propylene–vinyl chloride copolymers,[109] acrylonitrile–vinylidene chloride copolymer,[109] and phenyl polymers.[119]

The use of laser pyrolysis–gas chromatography–mass spectrometry (LP–GC–MS) has been used by Merritt and Sacher to study the effects of environmental exposure and field service on materials.[71,120–122] One of the studies involved the effect of weathering on an epoxy–glass fiber composite. During the weathering the principal oxidation products detected were acetone and acetophenone with the relative concentrations of these species increasing with exposure. The concentrations of other pyrolysis products such as toluene, benzene, and 2-phenyl-propane also increased with sample exposure time. This, together with the observation that no volatile organics could be detected from the unexposed material, suggested that outdoor exposure causes a lowering of the total cross-link density, thus increasing the polymer yield. The results were substantiated by dynamic mechanical measurements.[120]

Probe- and laser-type pyrolysis processes are the only ones that lend themselves to direct interfacing with the gas chromatograph. This is due to the fact that this pyrolysis process is very fast and the product mixture enters the column as a small slug. This is not the case for other much slower methods of thermal degradation in which trapping techniques are usually employed. Conley followed the combustion of poly(phenylene ether) polymers by trapping the effluents from the TGA and analyzing the trapped fraction and residues by GC and GC–MS.[43] and Chaigneau[123] and Einhorn and coworkers[102] followed the thermal degradation of polyurethanes by TGA–trapping–GC–MS. Einhorn used a low-temperature porous polymer trap to absorb the effluent gases from a Mettler thermoanalyzer. The trap was then heated and desorbed into a gas chromatograph where the oven temperature was held constant at −20° C. At this temperature most of the volatile materials collected at the head of the column. Upon completion of the trap desorption the gas chromatograph was temperature programmed for the analysis. Both EI and CI mass spectrometry were used in the analysis as well as a technique pioneered by Futrell of simultaneous EI–CI mass spectrometry.[102,124] This was accomplished by an AEI-MS-30 dual-beam double-focusing mass spectrometer. The instrument is modified such that one beam operates under CI condition and the other under EI condition. Thus when a sample elutes from the GC it is simultaneously analyzed by both beams of the mass spectrometer and both the EI and CI spectra are obtained. The utility of chemical ionization was demonstrated by its ability to successfully identify a series of amines produced from the degradation. The identification of amines by electron impact mass spectra is very difficult because they show extensive fragmentation, usually no molecular ions are observed, and amines of similar structure give almost identical spectra. CI–MS also has the ability to distinguish individual compounds in unresolved GC peaks. Einhorn and coworkers have successfully applied these techniques to the study of thermal degradation products and mechanisms for a number of polymeric systems under a variety of degradation conditions including TGA under vacuum, inert atmosphere[(125)] and oxygen-containing environments both in the flaming[(102)] and nonflaming mode[(102,125)] using tube furnaces,[(38,126)] reaction chambers[(102)] and the Aminco–NBS chamber[(127)] developed at the National Bureau of Standards for smoke density measurements.[(128)]

These techniques have also been applied by Einhorn and others to the monitoring of environment during animal exposures to the products of thermal degradation of polymers. During the course of animal exposure experiments conducted in the Aminco–NBS chamber, Petajan *et al.*[(128)] observed adverse neurological affects in rats exposed to the nonflaming combustion products from a fire-retarded urethane foam. Voorhees and coworkers[(126)] combusted the urethane in a tube-type reactor in air and identified the toxicant as a bicyclophosphate ester produced by reaction between the degradation products of the urethane and the degradation products of the phosphate fire retardant.

Woolley[(39,129,130)] studied the pyrolysis and combustion of natural and synthetic polymeric materials by a number of techniques. He has reported on a single- and double-stage tube furnace where the polymer is introduced into the primary tube furnace at temperatures up to 1000°C in air or nitrogen atmospheres. The effluent gases may be trapped and subsequently analyzed after the primary tube furnace or, alternatively, they may pass through a second furnace and undergo further high-temperature gaseous decomposition before analysis. This double-tube furnace system is used to model a full-scale fire. To obtain data on full-scale fires they used a special compartment–corridor facility 12.6 m long which was designed to represent a room attached to a long corridor (escape route). The compartment was fully instrumented and gas samples were extracted from various points in the test facility for analysis. The facilities have been used to study the synthetic polymers PVC,[(129,131)] polyurethane,[(39,130,132)] polyester foams,[(39)] and phenol formaldehyde laminates.[(133)] In the full-scale test, fuel loading ranged from domestic loads as small as two cushions to industrial loads up to 285 kg in a single burn.

The double-zone tube furnace experiments by both Woolley[(39)] and Israel *et al.*[(38)] indicated that when the degradation products from the decomposition of a polymeric material are passed through a second high-temperature zone, extensive secondary degradation occurs.

The mechanism of the thermal decomposition of PVC is one that has generated much interest and controversy. Whether HCl evolution precedes the evolution of benzene or the two evolve simultaneously has been discussed previously. The mechanism of hydrocarbon formation, primarily benzene, generated much controversy until the publication of an elegant study by O'Mara.[134] The question addressed was whether the formation of benzene took place by an intramolecular cyclization of the dehydrochlorinated polyene chain, or did it take place through a Diels–Alder-type condensation between neighboring polyene chains as proposed by Razuvaev.[135] The intermolecular mechanism received support from the observed cross-linking and increase in molecular weight during the early stages of the decomposition. To distinguish between these two mechanisms O'Mara synthesized poly(d_3-vinyl chloride) and subjected a 50:50 mixture of deuterated and undeuterated PVC to thermal degradation. If the intramolecular mechanism is in operation, then only fully deuterated and undeuterated benzene should be produced. If the intermolecular mechanism is in operation, partially deuterated benzene should predominate. The deuterium substitution is easily distinguished by mass spectrometry as each deutero atom adds 1 amu to the weight of the benzene molecule. The degradation products were analyzed by GC–MS and only fully deuterated and undeuterated benzene were observed, thus supporting the intramolecular mechanism.

Historically, the analysis of thermal degradation products has focused primarily on the volatile materials released. For the majority of these studies, a compound is considered volatile if it is analyzable by GC techniques. Very little attention has been paid to the high-boiling and solid residues except to determine their elemental composition. With the advent of high-pressure liquid chromatography (HPLC) it is now possible to analyze the high-boiling products as well. HPLC is a fractionation technique which is dependent upon the solubility of a compound and not on its volatility. At the present time there is one commercial interface available between a HPLC and mass spectrometer but this interface has seen only limited use. The few reported applications of HPLC to polymer degradation products involve manually collecting the individual fractions from the HPLC, evaporating off the solvent, and then performing MS analysis by direct inlet,[102] or in the case of the determination of polynuclear aromatic hydrocarbons in PVC smoke particulates[136] by GC–MS. Liao and Browner[136] developed a chromatographic method for the analysis of polynuclear aromatic hydrocarbons in PVC smoke particulates under simulated fire conditions. They collected particulates from the flaming and nonflaming combustion of PVC in the combustion products test chamber[137] at Georgia Tech. They identified about 25 different polynuclear aromatic hydrocarbons including the carcinogens benzo(*a*)pyrene and 7,12-dimethylbenz(*a*)anthracene.

3.1.3. Material "Fingerprinting" by Pyrolysis Mass Spectrometry

Direct pyrolysis–mass spectrometry was proposed as early as 1948[15,138] as a method of studying polymers. Each unique chemical system should, under thermal degradation, give rise to a unique mix of products. It would be expected that the mass spectrum of this product mix would be likewise unique. The direct identification of a sample should be possible from its pyrolysis–mass spectrum "fingerprint." This has been shown to be the case for pyrolysis–gas chromatography where the sample is pyrolyzed in the injection port of a gas chromatograph to yield a characteristic chromatographic spectrum of pyrolysis products (pyrogram). Groten[139] examined the pyrolysis of over 150 different polymers and found that almost all of them yielded unique pyrograms. The technique of pyrolysis–gas chromatography of polymers has recently been reviewed by Berezkin.[140] It must be pointed out that this technique has met with great skepticism due to the lack of interlaboratory reproducibility.[141]

Fingerprinting by direct pyrolysis–mass spectrometry has distinct advantages over chromatographic techniques. The superior sensitivity of the mass spectrometer allows for the use of samples in the microgram range. The ion source is under vacuum and thus the combination of vacuum conditions and very small sample size allows for rapid escape of the pyrolysis products from the pyrolysis zone, thus making the probability of secondary reactions between the products very low. This method has an additional advantage in its speed of analysis and ease of data handling. The analysis is complete within 1 or 2 min as opposed to chromatographic techniques, which require 1 or 2 hr per sample.

The direct pyrolysis in the normal electron impact ion source has been performed for many varied types of samples but the interpretation of the resulting spectra has been very difficult. This is due to the fact that each species produced by the thermal degradation undergoes mass spectral fragmentation in the ion source. Hughes[142] utilized a Curie-point pyrolyzer to examine a large number of materials of forensic interest including alkyd and acrylic paint flakes, synthetic and natural fibers, adhesives, greases, etc. Samples of 5 μg were pyrolyzed into an empty glass column and introduced into a normal electron impact ion source through a jet separator. The MS was repetitively scanned and at least 40 sequential scans were computer averaged to yield a composite mass pyrogram. The resulting mass pyrograms were usually characteristic of the class of sample and in many cases allowed for the discrimination of samples of similar composition within a class.

In order to simplify the mass pyrogram and insure that individual peaks have a more direct correlation to the individual products of the thermal degradation, it is necessary to minimize, or preferably eliminate, the fragmentation due to ionization in the mass spectrometer. There are several means available that will accomplish this: low-voltage electron impact ionization, field ionization–field desorption, and chemical ionization.

Low-voltage electron impact ionization has the advantage that it yields highly characteristic spectra with prominent molecular ions and limited fragmentation. Meuzelaar *et al.*[143,144] studied the direct Curie-point pyrolysis–low-voltage electron impact mass spectrometry of model phenolic polymers. The mass pyrograms from the phenolic polymers and from phenolic systems of biological origin were complex but similar, and in most cases it was possible to distinguish between the systems based on the differences in their spectra. Blasius[145] characterized the structure of a series of ion exchange resins. He was able to identify the type of network and functional groups present by the fingerprint spectra obtained by pyrolysis–low-voltage EI mass spectrometry and data obtained from pyrolysis–gas chromatography. Luderwald *et al.*[61,62] found that the application of pyrolysis–low-voltage EI mass spectrometry allowed them to differentiate between the thermal degradation and electron impact fragmentation and to establish the mechanism of both processes for a series of homo- and copolyamides. By the use of low-molecular-weight model compounds, the electron impact fragmentation mechanisms were established and peaks that were unique to each species found. By these peaks it was possible to differentiate between the piperazine polyamides of truxillic acid (head-to-head polymer) and truxinic acid (head-to-tail polymer) and to make quantitative correlation between the mass spectrum peaks and the amounts of each comonomer in the pyrolysis of some copolyamides of truxillic and adipic acid and pierazine. Both the truxillic and the truxinic acid can exist as *cis* or *trans* isomers. An attempt was made to distinguish between these isomers of the truxillic and truxinic polyamides. This was accomplished by temperature-programming the probe at a rate of 40°C/min with the mass spectrum taken every 10°C. The relative sum intensities are plotted versus temperature. From these MTA curves the initial temperature of decomposition and the peak maxima could be determined. The values were interpreted as a measure of the thermal stability of the systems and were used to distinguish between the different isomers.

Kistemaker *et al.*[37] combined laser pyrolysis and low-voltage EI mass spectrometry to study a series of seven synthetic polymers. Among them were poly(tetrafluoroethene) (PTFE), poly(chlorotrifluoroethene) (Kel-f), polystyrene, poly(methyl methacrylate), polyethene, and polypropene. Butyl rubbers were also studied. They demonstrated the general applicability and potential of this technique not only for identification of polymeric materials, but for determination of structure as well. Usually a major drawback to low-voltage EI is the low-ionization cross section of organic compounds for low-energy electrons and the fact that some of the compounds in the mixture may have appearance potentials higher than the ionizing voltage. Kistemaker demonstrated how this could be used to advantage. In the pyrolysis of poly(methyl methacrylate) a molecular peak is found at *m/e* 56. This corresponds to a molecular formula C_3H_4O and may be identified as either methylketene or acrolein. Methylketene has an appearance potential below 10 eV and acrolein well above 10 eV. When the spectrum was determined at 10 eV and the peak at *m/e* 56 was present, it was identified as methylketene.

Schulten[146] has recently reported on the first Curie-point pyrolysis in the ion source of a high-resolution field ionization mass spectrometer. In these experiments he pyrolyzed polysaccharides and observed almost exclusively molecular ions arising from the pyrolysis products.

A major obstacle encountered in the fingerprinting of materials is the problem of quantitatively determining when two spectra, or pyrograms, or fingerprints are different. This is not a trivial problem as the differences may be very subtle (e.g., relative abundances ratio of two peaks) and still contain a great deal of information. The application of computer-based pattern recognition techniques to this field should raise the sophistication by which we can distinguish structurally similar systems. These techniques are beginning to be applied to the pyrolysis of biological systems.[147]

At the present time the techniques of field desorption and chemical ionization have not been applied to the fingerprinting of synthetic polymers. Both of these techniques hold much promise for future studies.

3.2. *Dielectric Breakdown*

The degradation of polymers subjected to stress has been extensively studied in the case of thermal stress, but very little work has been done on the study of the chemical mechanism and evolution of volatile materials from polymers subject to other types of stress. The electrical insulating properties of many commercial polymers are well known and these properties are commercially exploited in the electrical industry. Materials in use as insulation are subject to electrical stress and undergo a degradation known as dielectric breakdown.

Smith and co-workers[148] studied the dielectric breakdown of the DuPont polyimide ML resin. The polyimide was coated on a copper wire and placed in a specially constructed reaction vessel. This reaction vessel was interfaced directly to the ion source of a mass spectrometer. The spectrum was recorded at room temperature at 0 and 1000 V; the temperature was then raised to 100° C and the spectrum taken first at 0 V and then again at 1000 V. The analysis was repeated at 200 and 300° C. The combination of thermal and electrical stress yielded a different mass spectrometric pattern than did the degradation under thermal stress alone. The most significant difference was the increased intensity of the water peak upon application of the electrical stress. The author speculated about possible mechanistic interpretations to explain the differences in the spectra.

3.3. Mechanical Stress-Induced Degradation

When a material is subjected to a mechanical stress it undergoes a stress-induced bond rupture. In extreme cases visible fracture of the material occurs. The study of stress-induced bond rupture of polymers has received increased attention and the application of mass spectrometry to follow the course of the degradation and to analyze the volatile degradation products was first suggested in 1966 by Regel.[149,150] He has since published a series of reports on the kinetics, activation energy, and volatile product analysis of polymers under load.[151] Polystyrene,[150,152] poly(vinyl alcohol),[150] poly(methyl methacrylate),[150,152] poly(vinyl toluene),[152] and polyethylene[152] were subjected to stress at the ion source of the mass spectrometer. The decomposition was found to begin immediately upon the application of the load. The rate of decomposition was proportional to the applied load. The kinetics of the decomposition under stress were correlated with the kinetics of thermal deformation. Dubrovskii *et al.*[153] studied the mechanical degradation of poly(tetrafluoroethylene), polycaprolactam, and poly(methyl methacrylate) during polymer–metal friction. They used an off-line technique to analyze the starting material and the polymer layer transferred to and adhering to the surface of a steel plate immediately after polymer–steel friction experiments. They could detect no differences in the spectra of the transferred layer and starting material.

Grayson and co-workers analyzed the volatile products produced by polymeric materials subjected to mechanical stress.[154–156] The ion source of a TOF mass spectrometer was modified to include a cell for stressing the polymers. The authors describe four separate cells that could apply stress in different modes. The first cell was used to apply tension to the point of fracture to a dog-bone-shaped sample. The second cell was a tensile apparatus with a sliding weight mechanism providing the force. The third was designed to stress and fracture filaments. The fourth cell consisted of a circular saw blade that could be used to abrade or saw materials such as composites. Polystyrene,[155] nylon,[156] and an epoxy composite[156] have been analyzed by these techniques. The authors found it was necessary to perform a separate analysis of the volatile material entrained in the polymer sample, e.g., solvent, unreacted monomer, etc., by a GC–MS technique in order to distinguish those materials from the degradation products.

3.4. Photolytic Degradation

The photolytic reactions of organic molecules and polymers are well known and have received much attention. The analysis of the photolytic decomposition of polymers directly by a mass spectrometric technique has been reported by Gorodetskii.[157] He described a photolysis apparatus that allowed the irradiation of polymer films in the vacuum ultraviolet region. The photolysis cell was interfaced directly to the mass spectrometer ion source. The sample could be irradiated as a film cast on a metal surface or as a strip of film. The strip of film was wound from one spindle to another, passing the sample through the photolysis beam. The mass spectrometer was continuously scanned during the exposure, thereby monitoring any evolved gases. The system also provided for the determination of quantum yield of the photolysis reactions.

4. Conclusion

The application of mass spectrometric techniques to the analysis of polymers and polymer degradation products cuts across the field of polymer chemistry. The breadth of application encompasses the range from direct analysis to degradation and reaction kinetics. This technique is now about 15 years old and its wide range of application was recognized in the earliest reports, but it is only during the last few years that we have seen a significant increase in publications reporting the use of this technique. This increase is due in large part to the development and increased availability of modern, computerized mass spectrometers designed for routine operation.

The field is very young and the technique has yet to be utilized and applied to many areas where its advantages are obvious. With the increased availability of GC–MS–DS systems and the further development of new ionizing methods, it is expected that this field will experience an extremely rapid growth.

5. References

1. M. S. B. Munson and F. H. Field, *J. Am. Chem. Soc.* **88**, 2621 (1966).
2. B. Munson, *Anal. Chem.* **49**, 772A (1977).
3. W. J. Richter and H. Schwarz, *Angew. Chem. Int. Ed.* **17**, 424 (1978).
4. M. G. Inghram and R. Gomer, *Z. Naturforsch.* **10a**, 863 (1955).
5. H. D. Beckey, *Naturwissenschaften* **45**, 259 (1958).
6. H. D. Beckey, *Angew. Chem. Int. Ed.* **8**, 623 (1969).
7. H. D. Beckey, *Field Ionization Mass Spectrometry*, Pergamon Press, Oxford (1971).
8. H. D. Beckey and H. R. Schulten, *Angew. Chem. Int. Ed.* **14**, 403 (1975).
9. W. V. Ligon, Jr., *Anal. Chem.* **50**(8), 1228 (1978).
10. A. K. Lee and R. D. Sedgwick, *J. Polym. Sci. Polym. Chem. Ed.* **16**, 685 (1978).
11. A. K. Lee and R. D. Sedgwick, *J. Polym. Sci. Polym. Chem. Ed.* **16**, 997 (1978).
12. J. R. Wright and T. J. Klingen, *J. Inorg. Nucl. Chem.* **32**, 2853 (1970).
13. S. Cucinella, T. Salvatori, C. Busetto, G. Perego, and A. Mazzei, *J. Organomet. Chem.* **78**(2), 185 (1974).
14. R. P. Lattimer, K. R. Welch, J. B. Pausch, and V. Rapp, Varian Mat Application Note No. 27, Varian Mat GmbH, Bremen, West Germany (1978).

15. S. L. Madorsky and S. Strauss, *Ind. Eng. Chem.* **40**, 848 (1948).
16. L. A. Wall, *J. Res. Natl. Bur. Stand.* **41**, 315 (1948).
17. P. D. Zemany, *Anal. Chem.* **24**, 1709 (1952).
18. D. L. Harms, *Anal. Chem.* **25**, 1140 (1953).
19. R. L. Levy, *J. Gas Chromatogr.* **5**, 107 (1967).
20. R. L. Levy, *Chromatog. Rev.* **8**, 48 (1966).
21. R. L. Levy and D. L. Fanter, *Anal. Chem.* **41**, 1465 (1969).
22. R. L. Levy, D. L. Fanter, and C. J. Wolf, *Anal. Chem.* **44**, 38 (1972).
23. M. Krejici and M. Deml, *Collect. Czech. Chem. Commun.* **30**, 3071 (1965).
24. W. Simon and H. Giacobbo, *Chem. Ing. Tech.* **37**, 709 (1965).
25. Ch. Buhler and W. Simon, *J. Chromatogr. Sci.* **8**, 323 (1970).
26. E. M. Andersson and I. Ericsson, in *Analytical Pyrolysis, Proceedings of the 3rd International Symposium on Analytical Pyrolysis*, Amsterdam, 1976 (C. E. Roland Jones and C. A. Cramers, eds.), p. 415, Elsevier, Amsterdam (1977).
27. N. E. Vanderborgh, in *Analytical Pyrolysis, Proceedings on the 3rd International Symposium on Analytical Pyrolysis*, Amsterdam, 1976 (C. E. Roland Jones and C. A. Cramers, eds.), p. 235, Elsevier, Amsterdam (1977).
28. N. E. Vanderborgh and W. T. Ristau, *Am. Lab.*, 41 (May 1973).
29. D. L. Fanter, R. L. Levy, and C. J. Wolf, *Anal. Chem.* **44**, 43 (1972).
30. W. T. Ristau and N. E. Vanderborgh, *Anal. Chem.* **44**, 359 (1972).
31. B. T. Guran, R. J. O'Brien, and D. H. Anderson, *Anal. Chem.* **43**, 702 (1971).
32. W. K. Joy, in *Time-of-Flight Mass Spectrometry* (D. Price and J. E. Williams, eds.), p. 45, Pergamon Press, London (1969).
33. W. K. Joy and B. G. Reuben, *Dynamic Mass Spectrometry* (D. Price and J. E. Williams, eds.) Vol. 1, p. 183, Heyden & Stone, London (1970).
34. N. E. Vanderborgh and W. T. Ristau, *J. Chromatogr. Sci.* **11**, 535 (1973).
35. N. E. Vanderborgh and W. T. Ristau, *Anal. Chem.* **45**, 1529 (1973).
36. J. Q. Walker, *Chromatographia* **5**, 547 (1972).
37. P. G. Kistemaker, A. J. H. Boerboom, and H. L. C. Meuzelaar, *Dyn. Mass Spectrom.* **4**, 139 (1975).
38. S. C. Israel, D. A. Chatfield, and I. N. Einhorn, unpublished results.
39. D. J. Woolley and P. J. Fardell, in Proceedings of the International Symposium on Toxicity and Physiology of Combustion Products, University of Utah, March 1976.
40. H. G. Langer and R. S. Gohlke, *Anal. Chem.* **35**, 1301 (1963).
41. H. L. Friedman, ML-TDR-64-274, Air Force Materials Laboratory Wright-Patterson Air Force Base, Dayton, Ohio, August 1964.
42. H. L. Friedman, *J. Macromol. Sci. (Chem.)* **A1**(1), 57 (1967).
43. G. P. Shulman, *J. Macromol. Sci. (Chem.)* **A1**(1), 107 (1967).
44. E. Jacobi, I. Luderwald, and R. C. Schultz, *Makromol. Chem.* **179**(1), 277 (1978).
45. G. J. Mol, *Thermochim. Acta.* **10**(3), 259 (1974).
46. C. Z. Carroll-Porezynski, *Composites* **4**(1), 9 (1973).
47. G. A. Kleineberg, D. L. Geiger and W. T. Gormley, *Makromol. Chem.* **175**(2), 483 (1974).
48. R. Sakamoto, T. Ozawa, and M. Kanazashi, *Thermochim. Acta.* **3**(4), 291 (1972).
49. O. P. Korobeinichev, A. S. Schmelev, V. G. Voronov, and G. I. Anisiforov, *Therm. Anal. Proc. Int. Conf.*, 4th, Budapest, Jl 8-13, 1974. (I. Buzas, ed.), p. 77, Heyden, London (1975).
50. H. L. Friedman, G. A. Griffith, J. R. Mallin, and N. M. Jaffe, *Thermochim. Acta.* **8**(1-2), 119 (1974).
51. R. W. Mickelson, K. J. Voorhees, and P. W. Ran, in Polymer Conference Series, University of Utah, July 1975.
52. F. Zitomer, *Anal. Chem.* **40**, 1091 (1968).
53. I. A. Murdoch and L. J. Rigby, in *Dynamic Mass Spectrometry* (D. Price, ed.), Vol. 3, p. 244, Heyden, London (1972).

54. M. Ryska, J. Babka, and M. Kolinsky, *Sb. Prednasek "Makrotest 1973"* **1**, 243 (1973).
55. R. H. Wiley and L. H. Smithson, *J. Macromol. Sci.* **A2**, 589 (1968).
56. R. H. Wiley, *Trans. N.Y. Acad. Sci.* **32**, No. 6 (1970).
57. R. H. Wiley, *Polym. Prep.* **11**, 747 (1970).
58. R. H. Wiley, *J. Polym. Sci. A-1* **9**(1), 129 (1971).
59. I. Luderwald and H. Urrutia, in *Analytical Pyrolysis, Proceedings of the 3rd International Symposium*, Amsterdam, 1976 (C. E. Roland Jones and C. A. Cramers, eds.), p. 139, Elsevier, Amsterdam (1977).
60. F. Bottino, S. Foti, G. Montaudo, S. Pappalardo, I. Luderwald, and M. Przybylski, *J. Polym. Sci. Polym. Chem. Ed.* **16**, 3131 (1978).
61. I. Luderwald, M. Przybylski, H. Ringsdorf, S. Foti, and G. Montaudo, in *Analytical Pyrolysis, Proceedings of the 3rd International Symposium*, Amsterdam, 1976 (C. E. Roland Jones and C. A. Cramers, eds.), p. 297. Elsevier, Amsterdam (1977).
62. S. Caccamese, S. Foti, P. Maravigna, G. Montaudo, A. Recca, I. Luderwald, and M. Przybylski, *J. Polym. Sci. Polym. Chem. Ed.* **15**, 5 (1977).
63. I. Luderwald and H. R. Kricheldorf, *Angew. Makrmol. Chem,.* **56**(1), 1973 (1976).
64. C. O'Oyly-Watkins, in Proceedings of the TTCP-3 Critical Reviews; Techniques for the Characterization of Polymeric Materials, Watertown, Massachusetts (January 1977), p. 197.
65. C. Reed, *Brit. Polym. J.* **6**, 1 (1974).
66. H. D. Stenzenberger, K. U. Heinen, and D. O. Hummel, *J. Polym. Sci. Polym. Chem. Ed.* **14**, 2911 (1976).
67. R. M. Lum, *J. Appl. Polym. Sci.* **20**, 1635 (1976).
68. R. M. Lum, *Thermochim. Acta* **18**(1), 73 (1977).
69. R. M. Lum, *J. Polym. Sci. Polym. Chem. Ed.* **15**, 489 (1977).
70. K. R. Thompson, *Anal. Chem.* **48**(4), 696 (1976).
71. R. E. Sacher, in Proceedings of the TTCP-3 Critical Review, Techniques for the Characterization of Polymeric Materials, Watertown, Massachusetts (1977), p. 319.
72. A. T. DiBenedetto and D. A. Scola, *J. Colloid. Interface Sci.* **64**(3), 480 (1978).
73. G. R. Sparrow and H. E. Mishmash, ASTM Spec. Tech. Publ. 1977, A643, *Quant. Surf. Anal. Mater.*, 164 (1978).
74. G. D. Tantsyrev, G. D. Povolotskaya, and N. A. Kleimenov, *Vysokomol. Soedin. Ser. A* **19**(9), 2361 (1977).
75. V. A. Lavrenko, A. G. Koval, Yu. A. Klimovskii, A. A. Lebedev, R. N. Lyanikova, and I. N. Frantsevich, *Dokl. Akad. Nauk SSSR*, **226**(1), 122 (1976).
76. G. D. Tantsyrev, N. A. Kleimenov, M. I. Povolotskaya, and N. M. Bravaya, *Vysokomol. Soedin. Ser. A* **18**(10), 2218 (1976).
77. G. D. Tantsyrev and M. I. Povolotskaya, *Khim. Vys. Energ.* **9**(4), 380 (1975).
78. M. Aoki, K. Kurosaki, *Kobunshi Kagako* **29**(6), 442 (1972).
79. W. L. Baun, *Pap. Meet. Am. Chem. Soc. Div. Org. Coat. Plast. Chem.* **36**(1), 344 (1976).
80. R. M. Fristrom and C. Grunfelder, FPP TR 25 APL Fire Problems Program, Applied Physics Laboratory, The Johns Hopkins University, Silver Spring, Maryland (1974).
81. R. M. Fristrom, *J. Fire Flammability* **5**, 289 (1974).
82. R. M. Fristrom, C. Grunfelder, and L. W. Hunter, *Combust. Flame* **27**(1), 33 (1976).
83. H. Schacke, L. W. Hunter, R. M. Fristrom, and C. Grunfelder, 16th Combustion Symposium, Combustion Institute, Pittsburgh, Pennsylvania (1976).
84. R. M. Fristrom, *Int. J. Mass. Spectrom. Ion. Phys.* **16**(1-2), 15 (1975).
85. I. C. McNeill, *J. Polym. Sci. A-1* **4**, 2479 (1966).
86. I. C. McNeill, *Eur. Polym. J.* **3**, 409 (1967).
87. I. C. NcNeill and D. Neil, *Thermal Analysis* (R. F. Schwenken and P. D. Garn, eds.), p. 353 Academic Press, New York (1969).
88. I. C. McNeill, L. Ackerman, S. N. Gupta, M. Zulfiqar, and S. Zulfiqar, *J. Polym. Sci. Polym. Chem. Ed. 15*, 2381 (1977).

89. G. Camino, N. Grassie, and I. C. McNeill, *J. Polym. Sci. Polym. Chem. Ed.* **16**, 95 (1978).
90. A. Jamieson and I. C. McNeill, *J. Polym. Sci. Polym. Chem. Ed.* **14**, 1839 (1976).
91. A. Jamieson and I. C. McNeill, *J. Polym. Sci. Polym. Chem. Ed.* **16**, 2225 (1978).
92. R. H. Still, M. B. Evans, and A. Whitehead, *J. Appl. Polym. Sci.* **16**(12), 3207 (1972).
93. R. H. Still and A. Whitehead, *J. Appl. Polym. Sci.* **16**(12), 3223 (1972).
94. J. Chiu and A. J. Beattle, *Thermochim. Acta* **21**(2), 263 (1977).
95. I. C. McNeill, L. Ackerman, and S. N. Gupta, *J. Polym. Sci. Polym. Chem. Ed.* **16**, 2169 (1978).
96. L. Ackerman and W. J. McGill, *J.S. Afr. Chem. Inst.* **26**, 82 (1973).
97. D. O. Hummel, H. D. Schueddemage, and K. Ruebenacker, *Monogr. Mod. Chem.* **6**, 355 (1974).
98. J. Mitchell, Jr. and J. Chiu, *Anal. Chem.* **45**(3), 273R (1973).
99. K. Hiramatsu, *Kobunshi* **17**(200), 1052 (1968).
100. W. Linder, *J. Chromatogr.* **151**(3), 406 (1978).
101. G. F. Sykes and P. R. Young, *J. Appl. Polym. Sci.* **21**(9), 2393 (1977).
102. I. N. Einhorn, D. A. Chatfield, K. J. Voorhees, F. D. Hileman, R. W. Mickelson, S. C. Israel, J. H. Futrell, and P. W. Ryan, *Fire Res.* **1**, 41 (1977).
103. F. D. Hileman, K. J. Voorhees, L. H. Wojcik, M. M. Birky, P. W. Ryan, and I. N. Einhorn, *J. Polym. Sci. Polym. Chem. Ed.* **13**, 571 (1975).
104. K. J. Voorhees, in Proceedings of the International Symposium on Toxicity and Physiology of Combustion Products, University of Utah, March 1976.
105. K. J. Voorhees, F. D. Hileman, I. N. Einhorn, and J. H. Futrell, *J. Polym. Sci. Polym. Chem. Ed.* **16**, 213 (1978).
106. R. J. Gritter, M. Seeger, and E. Gipstein, *J. Polym. Sci. Polym. Chem. Ed.* **16**, 353 (1978).
107. R. J. Gritter, M. Seeger, and D. E. Johnson, *J. Polym. Sci. Polym. Chem. Ed.* **16** 169 (1978).
108. N. Grassie, *Pure Appl. Chem.* **30**, 119 (1972).
109. J. Mitera, *Sb. Vys. Sk. Chem. Technol. Praze Technol. Paliv.* **D33**, 123 (1976).
110. M. M. O'Mara, *J. Polym. Sci. A-1* **8**, 1887 (1970).
111. G. Oehme, H. Baudisch, and H. Mix, *Makromol. Chem.* **177**(9), 2657 (1976).
112. M. Przybylski, H. Ringsdorf, and H. Ritter, *Makromol. Chem.*, Suppl. **1**, 29 (1975).
113. A. Ninagawa, Y. Katsuya, H. Matsuda, and S. Matsuda, *Makromol. Chem.* **174**, 225 (1973).
114. J. L. Cotter, *Org. Mass. Spectrom.* **7**(1), 11 (1973).
115. J. L. Cotter, U.S. Nat. Tech. Inform. Serv., AD Rep. 1972, No. 746730, NTIS, *Gov. Rep. Announce (U.S.)* **72**(19), 50 (1972).
116. R. L. Foltz, M. D. Neher, and E. R. Hinenkamp, *Appl. Polym. Symp.*, No. 10, 195 (1969).
117. N. A. Nikolaeva, B. G. Belenkis, N. A. Glukhov, Yu. V. Zhuravlev, and Yu. N. Sazanou, *Vysokomol. Soedin. Ser. A* **12**(11), 2625 (1970).
118. N. Sellier, C. E. R. Jones, and G. Guiochon, *J. Chromatogr. Sci.* **13**(8), 383 (1975).
119. M. T. Jackson, Jr., and J. Q. Walker, *Anal. Chem.* **43**, 74 (1971).
120. C. Merritt, Jr., R. E. Sacher, and B. A. Petersen, *J. Chromatog.* **99**, 301 (1974).
121. C. Merritt, Jr., M. L. Bazinet, R. E. Sacher, and J. Sprouse, in *Analytical Pyrolysis, Proceedings of the 3rd International Symposium on Analytical Pyrolysis*, Amsterdam, 1976 (C. E. Roland Jones and C. A. Cramers, eds.), p. 419, Elsevier, Amsterdam (1977).
122. C. Merritt, Jr., in *Analytical Pyrolysis, Proceedings of the 3rd International Symposium on Analytical Pyrolysis*, Amsterdam, 1976 (C. E. Roland Jones and C. A. Cramers, eds.), p. 211, Elsevier, Amsterdam (1977).
123. M. Chaigneau and G. LeMoan, *Ann. Pharm. Fr.* **30**(6), 409 (1972).
124. J. H. Futrell, U.S. NTIS, PB Rep. PB-259108 (1975), NTIS, *Gov. Rep. Announce (U.S.)* **77**(4), 67 (1977).
125. I. N. Einhorn, D. A. Chatfield, and D. J. Wendel, in *Analytical Pyrolysis, Proceedings of the 3rd International Symposium on Analytical Pyrolysis*, Amsterdam, 1976 (C. E. Roland Jones and C. A. Cramers, eds.), p. 421, Elsevier, Amsterdam (1977).

126. K. J. Voorhees, I. N. Einhorn, F. D. Hileman, and L. H. Wojcik, *Polym. Lett.* **13**, 293 (1975).
127. AMINCO-NBS Smoke Density Chamber, American Instrument Corp., Silver Spring, Maryland, D. Gross, J. J. Loftus, and A. F. Robertson ASTM Special Technical Publication Number 422 (1967).
128. J. H. Petajan, K. J. Voorhees, S. C. Packham, R. C. Baldwin, I. N. Einhorn, M. L. Grunnet, B. G. Dinger, and M. M. Birky, *Science* **187**, 742 (1975).
129. W. D. Woolley, *Brit. Polym. J.* **3**(4), 186 (1971).
130. W. D. Woolley, *Brit. Polym. J.* **4**(1), 27 (1972).
131. W. D. Woolley, *Plast. Polym.* **40**, 203 (1972).
132. W. D. Woolley, *J. Fire Flammability Combust. Toxicol. Suppl.* **1**, 259 (1974).
133. W. D. Woolley, Fire Research Station Fire Research Notes, 851, 852 Borehamwood, United Kingdom (1970).
134. M. M. O'Mara, in Proceedings of the International Symposium on Toxicity and Physiology of Combustion Products, University of Utah, March 1976.
135. G. A. Razuvaev, *Dokl. Akad. Nauk SSSR* **170**(6), 1342 (1966).
136. J. C. Liao and R. F. Browner, *Anal. Chem.* **50**, 1683 (1978).
137. C. P. Bankston, R. A. Cassanova, E. A. Powell, and B. T. Zinn, *J. Fire Flammability* **7**, 165 (1976).
138. L. A. Wall, *J. Res. Natl. Bur. Stand.* **41**, 315 (1948).
139. B. Groten, *Anal. Chem.* **36**, 1206 (1964).
140. V. G. Berezkin, V. R. Alishoyev, and I. B. Nemirovskaya, *Gas Chromatography of Polymers, Journal of Chromatography Library*, Vol. 10, Chap. 7, Elsevier, Amsterdam (1977).
141. N. B. Coupe, C. E. R. Jones, and S. G. Perry, *J. Chromatogr.* **47**, 291 (1970).
142. J. C. Hughes, B. B. Wheals, and M. J. Whitehouse, *Analyst (London)* **102**(1211), 143 (1977).
143. H. L. C. Meuzelaar, K. Haider, B. R. Nagar, and J. P. Martin, *Goederma* **17**, 239 (1977).
144. K. Haider, B. R. Nagar, C. Saiz, H. L. C. Meuzelaar, and J. P. Martin, Proc. FAO Conference, Braunschweig, West Germany (1976).
145. E. Blasius, H. Haeusler, and H. Lander, *Talanta* **23**(4), 301 (1976).
146. H. R. Schulten and W. Goertz, *Anal. Chem.* **50**(3), 428 (1978).
147. E. Eshuis, P. G. Kistemaker, and H. L. C. Meuzellar, in *Analytical Pyrolysis, Proceedings of the 3rd International Symposium on Analytical Pyrolysis*, Amsterdam, 1976 (C. E. Roland Jones and C. A. Cramers, ed.), p. 151, Elsevier, Amsterdam (1977).
148. S. R. Smith, S. A. S. Smith, J. F. Johnson, and J. Tanaka, *IEEE Trans. Electr. Insul.* **EI-12**(6), 442 (1977).
149. V. R. Regel, T. M. Muinov, and O. F. Pozdnyakov, in *Proceedings of the Physical Basis of Yield and Fracture Conference*, p. 194, Oxford University Press, Oxford (1966).
150. V. R. Regel and T. M. Muinov, *Vysokomolekul. Soedin* **8**(5), 841 (1966).
151. V. R. Regel, O. F. Pozdnyakov, and A. V. Amelin, *Mekh. Polim.* **1**, 16 (1975).
152. V. R. Regel, and O. F. Pozdnyakov, *Plaste Kautsch* **19**(2), 99 (1972).
153. V. S. Dubrovskii, A. I. Sviridenok, V. A. Smurugov, V. G. Savkin, and V. A. Belyi, Deposited Doc., VINITI, 173 (1975), avail. BLLD.
154. M. A. Grayson, R. L. Levy, D. L. Fanter, and C. J. Wolf, in *24th Annual Conference on Mass Spectrometry*, ASMS, San Diego, California (May 1976).
155. M. A. Grayson, C. J. Wolf, R. L. Levy, and D. B. Miller, *J. Polym. Sci. Polym. Phys. Ed.* **14**(9), 1601 (1976).
156. C. J. Wolf and M. A. Grayson, in Proceedings of the TTCP-3 Critical Review, Techniques for the Characterization of Polymeric Materials, Watertown, Massachusetts (1977), p. 117.
157. I. G. Gorodetskii, V. E. Skurat, V. L. Tai'Roze, and S. A. Yakovlev, *Tezisy Dikl. Vses. Konf. Spektrosk. Vak. Ul'Tra Fioleta. Vzaimodeistviyu Izluch. Veshchestvom* **4**, 20 (1975).

Index

MIX
Papier aus verantwortungsvollen Quellen
Paper from responsible sources
FSC® C105338

If you have any concerns about our products,
you can contact us on
ProductSafety@springernature.com

In case Publisher is established outside the EU,
the EU authorized representative is:
Springer Nature Customer Service Center GmbH
Europaplatz 3, 69115 Heidelberg, Germany

Printed by Libri Plureos GmbH
in Hamburg, Germany